What are the vanishing ships?

Liners, schooners, barks and freighters that vanish inexplicably. Submarines that never resurface and cannot be found. Doomed ghost ships that reappear for centuries.

Who are the missing men?

Entire crews dead for unknown causes. Drowned men who return to their ships. Sailors lost in bizarre circumstances.

What are the mysteries of the sea?

Death ships. Seaquakes. Phantom battles. Disasters. All without explanation. All documented. All true.

VANISHED - WITHOUT A TRACE!

BILL WISNER

With Sketches by Larry Kresek

A BERKLEY MEDALLION BOOK
published by
BERKLEY PUBLISHING CORPORATION

This is for my four girls,
all with stellar billing here:
Dorothy, Ruth, Judy and Elizabeth

CONTENTS

Introduction

"The sea" is a common collective term for all bodies of salt water, placid bays to mighty oceans. It represents a world whose immensity boggles the mind. The sea dwarfs continents in its coverage of 70.8 percent of our globe's surface. Its great deeps—six to seven miles down in some places—submerge huge mountain ranges strung out along ocean floors. The sea has always been our planet's most enormous frontier. What's more, since it is mostly a hidden, unexplored expanse, it defiantly promises to retain that frontier status for a long time to come. It may turn out that we become familiar with the moon's details before we learn even an appreciable amount of what there is to be known about the sea.

With its changing moods the sea is both a tantalizing wooer and a harsh master. (Make no mistake, it is still man's master.) In one mood it smiles in welcome and bounteously provides food, transportation routes and recreational pleasure. In another facet of temperament it turns on man in unbridled fury, taking away his possessions—and often his life. Its moods can change unpredictably and quickly, and foolish indeed are humans who think they can be certain of them.

Those are reasons for the sea's irresistibility. They also are reasons why man's ages-old love affair with the deep, far-flung water kingdom has spawned innumerable strange stories, and why more are in creation even as you read this. The chronicles parade through an almost limitless range. Some are a legacy of legends or old tales whose fabric is fancy interwoven with fact. Many are baffling mysteries, promising to remain forever unsolved. Others lie in the realm of the supernatural and defy explanation. And there's a large assortment of true horror stories, too, starkly real. Many are absolutely incredible. All are astounding in one way or another.

This book presents a wide selection of such sea stories. I trust that you will find them as fascinating to read as I did to write.

Bill Wisner
Brightwaters, New York

1

Presto! Gone!

There's a joke about a magician who was performing in the lounge of an ocean liner. Just as he built to the climax of his act, something disastrous happened deep in the ship's bowels and she disappeared beneath the waves in a matter of minutes. The magician and one passenger were the sole survivors and found themselves clinging to the same piece of flotsam. The passenger shook his head in amazement. "That, sir," he enthused to the magician, "is the greatest trick I've ever seen!"

In some respects the old canard isn't as far-fetched as it sounds.

Among the most fascinating of all enigmas in man's ages-old love affair with the sea are disappearances of entire ships—together with everyone aboard them, and no clues to their fate. Not one survivor to tell about the disaster, not even a corpse . . . not even a fragment of wreckage to provide a clue. Over the years there have been many such vanishing acts.

True, there would be a certain satisfaction in knowing how these disappearances came about. But then, let's face it, their attraction would pale considerably. After all, the most intriguing thing about a mystery is its lack of solution.

The "President" Vanishes

The scene at wharfside in New York that March day of 1841 was a prototype for many such tableaux to follow in decades ahead. A transatlantic passenger ship was about to depart for England. A festive mood infected travelers looking forward to a sea voyage and a vacation in Europe. There were some impatient people too, foreign visitors anxious to get home. And here and there one saw sorrow at parting. Aboard ship were 121 passengers, one of whom was prominent actor Tyrone Power, grandfather of the film star who was to bear his name. Mr. Power was on his way to a stage engagement in London.

Apprehension couldn't have been among the mixed emotions at dockside. Built in England and launched only two years earlier, the

President was widely considered one of the finest ships of her day. Rated as rugged and very seaworthy, and performance-proven on an ocean noted for its antisocial moods, she was manned by a veteran crew under the command of a skipper named Roberts.

Moreover, here was an exciting example of the latest in sea transportation. The *President* was powered by both engines and sails. In addition to the safety aspect that combination implied, there was no need to worry about wind failures that could becalm sailing ships for days. We can be sure that her accommodations and cuisine were of the finest, and that everything possible was done to cater to the creature comforts of her guests.

Amid farewells, the *President* eased away from her berth and soon was entering the roadstead of New York Harbor. Just beyond lay the open Atlantic; and far beyond the horizon, her destination, Liverpool. In between were 3,000 miles of pleasure—relaxation in tonic salt air, social functions at night, splendid menus (for those not overtaken by *mal de mer*), and opportunities to make new friends. The *President* should fetch Liverpool in sixteen days.

But she never arrived.

What Happened?

We'll never know for sure what befell the *President*. A clue of sorts may lie in the fact that on the day after she sailed a vicious sea storm lashed a segment of her course. Yet, when that weather information reached New York and Liverpool via other ships it caused no great concern for the *President*, since she was considered one of the safest, most seaworthy vessels. Besides, two smaller ships, *Orpheus* and the packet *Virginia,* had made it safely to Liverpool through the storm, suffering only delays. Furthermore, a London newspaper announced that the *Orpheus* departed New York after the *President* and overtook her at sea, which would have placed her beyond the storm area and presumably safe. (Later, the captain of the *Orpheus* was to deny he sighted the *President.*) The public wavered between optimism and fear, finding a measure of hope in the thought that the liner was just delayed and probably would appear in port before long. Perhaps she had experienced engine difficulties and was coming in under sail, in which case a delay could be expected.

But as days passed without sight or word of the *President,* apprehension overshadowed optimism. And when the steamer *Britannia*, previously reported long overdue with the liner, arrived late but safe in Liverpool, worry intensified. It was heightened further by reports of huge ice masses drifting in the Atlantic off Newfoundland. Finally, at the end of March, London newspapers issued an announcement that was tantamount to declaring the *President* lost. Obviously, she had met a fatal calamity somewhere out on the Atlantic . . . but what?

Logic accused the storm of doing her in, yet she had survived other weather just as bad, if not worse. Too, the ship's size, seaworthiness, and expert crew pointed against the possibility. Theories would have to include a boiler explosion, one that crippled her and left her helpless for the storm to finish off, or perhaps even a blast that disemboweled her.

The explosion theory had supporters. Although thirty-seven years had elapsed since Robert Fulton demonstrated the practicality of steam power for ships by chugging up and down New York's Hudson River in his *Clermont*, there remained strong suspicion about the safety of this form of propulsion. Perhaps there was justification in the *President*'s case; but, like other possible causes, it has to stay in the realm of conjecture. So, too, does a contemporary belief that she had a lethal encounter with an iceberg. This had some substance. Errant ice masses were a possibility in the North Atlantic at that time of year, and there were no iceberg patrols such as we now have. But nothing conclusive on that score came to light, either.

The *President*'s actual fate is unknown to this day. To all intents and purposes she vanished without a trace. Radio might have provided a clue, but there was none at that time. No wreckage or debris of any kind was found. Nor did any bodies or survivors ever turn up.

Sometime afterward, however, a strange postscript was announced. An Irish newspaper reported the recovery at sea of a bottle which purportedly carried a scrawled message reading: "The *President* is sinking. God help us all. Tyrone Power." This message was to become something of a mystery in itself, and a controversial one. In some quarters it was discredited outright as the brainchild of a publicity seeker, prankster or crackpot. Doubt of its genuineness persisted, yet it apparently was never definitely proved to be a hoax. Mrs. Power's reaction to it wasn't recorded. In any case, Tyrone Power didn't say why the ship was sinking (understandable, all things considered)—if, indeed, he did scribble the message. That secret went with him to Davy Jones's locker.

Loss of the *President* was followed by cruel hoaxes, a tangle of conflicting rumors, waterfront scuttlebutt, all kinds of theories, and even some outright fabrications.

One story was a pulse-boosting report that a special train had sped from Liverpool to Birmingham in central England with news that the *President* had arrived in port, badly battered but safe. This

3

rumor spread like a brush fire in a high wind, but turned out to be completely false. Then came published reports of letters allegedly written to their wives by the missing ship's skipper and Tyrone Power. Supposedly they came from the island of Madeira, out in the Atlantic, and in essence both said the *President* was safe in a harbor there. Captain Roberts' missive blamed his vessel's delay on forced repairs to her engine and rudder, presumably damaged by a storm. Mrs. Roberts later denied that she received such a letter. Friends of the missing actor pronounced the letter to his widow a forgery and hoax. Her comments weren't noted, but apparently she agreed with the friends' verdict.

Later Mrs. Roberts received a genuine communiqué, this from a former master of the *President*. He declared his belief that the missing steam-sail ship had experienced engine or rudder trouble. He intimated that she was progressing under sail and suggested that she had altered course for Bermuda, possibly to circumnavigate a region of adverse winds. He echoed the opinion of naval experts that the *President* was too sturdy to have been sunk. The letter undoubtedly gave Mrs. Roberts some comfort, but unhappily the correspondent hypotheses turned out to be wrong.

Before that was proved, however, there came another startler that seemed to support the former master's theory. It was word from Bristol, conveyed by letter to Captain Roberts' brother-in-law in London, that an unidentified vessel had put in at Waterford, a coastal hamlet in southern Ireland, with joyful news that the *President* was on her way home from Bermuda. This intelligence got around just as fast as jubilant human voices could distribute it. It was scotched abruptly by an authoritative sequel from Waterford officials who denied the whole thing, including the presence of a vessel that was supposed to have delivered the electrifying news. Also proved to be a heartbreaking dud was a newspaper item of May 1841, stating that a Portuguese vessel had sighted a steamship resembling the *President* far at sea, apparently disabled by engine trouble. This didn't ring true on several counts, not the least of which was the fact that the *President* could have progressed under sail.

Many years later a possible key to the mystery came to light with an exciting find on the Massachusetts coast. A Mr. John Blake, resident of a seaside village in the Bay State, was walking along a beach when he caught sight of a large piece of wood, nearly eight feet long and more than a foot wide, which had drifted in near shore. Mr. Blake managed to retrieve the plank when a wave tumbled it in among rocks just above the tide line.

Evidently he knew the story of the *President*, for his excitement

4

was great as he read the legend PRESIDENT spelled out on the plank in large, bold letters. Actually the word wasn't quite complete; the "T" and part of the "N" were missing where the plank had splintered. Originally—and this is purely my interjection—the name could have been PRESIDENTE, Spanish for "President." Or perhaps two words, such as PRESIDENT ADAMS, although subsequent study pointed to the name having been a single word.

It was thought that the plank may have been the letterboard of the type placed on either side of a vessel's pilothouse for identification at sea. And it was considered possible that it had come from the long-missing steamer. Experts pronounced its hardware of the vintage used in that liner's era, and its general appearance was that of wood immersed in sea water for a long time. The inference, of course, was that it had drifted around the ocean all those years. As has been proved by derelicts, that was quite possible.

But it didn't solve the enigma of the *President*.

The "Atalanta" Mystery

Midwinter is not an optimum time in which to sail across the Atlantic Ocean. Even in what passes for a reasonably good winter mood, it can be surly, and its hibernal storms are notorious. Nevertheless, mariners have been braving the Atlantic in all seasons for centuries, even in creaky little ships such as the immortal *Mayflower* and Chris Columbus' *Nina, Pinta,* and *Santa Maria*.

So there was no particular need for any worry among the 300-plus naval cadets and officers aboard the British Navy's well-built training ship H.M.S. *Atalanta* when she stood out from Bermuda on a voyage home in January 1880. With her sails bellying in the wind, Captain Francis Sterling estimated they would draw within sight of Spithead, England, on or about March 1st.

They didn't. Nor did they arrive by April 1st . . . or ever.

While allowances could be made for delays of sailing ships, when the *Atalanta* became two weeks overdue there was substantial cause for alarm. In response, the British Navy launched a far-flung sea search. Vessels fanned out across thousands of square miles of ocean to cover all possibilities. The search continued on into that spring, with completely negative results. No trace of the missing ship or anyone aboard her was found. Her disappearance was a clueless mystery.

Investigation

That spring and during the months following, the British Admiralty received and weighed all kinds of reports and rumors. Even desperation theories were given careful consideration. One was that the training vessel had been blown far off course. It didn't seem too likely, but it was explored. The possibility seemed to gain some support when the British Navy's *Avon* came upon fragments of an unidentified ship floating around off the Azores. The *Avon* probed the area, but found nothing to link the wreckage with the missing *Atalanta*.

In early June the master of a sailing vessel arriving at an English port told about sighting a raft and then some corpses dressed in what might have been the uniforms of naval cadets and/or officers, about two months earlier. Apparently he didn't pause to check, and his report became a loose end. Later in June there were reports of a bottle being retrieved from the sea not far off the Massachusetts coast and a piece of wood being found on a strand in Nova Scotia, both of which purportedly carried distress messages from the *Atalanta* and mentioned her going down. However, if we can judge by accounts of the findings there's an interesting discrepancy in their respective dates of the sinking. The bottle message said April 17; the fragment of wood was dated April 12. Perhaps that's why little credence seems to have been given to either. And an August report that the *Atalanta*'s figurehead (a prominent and often elaborate addition to the bow of a sailing ship) had been found subsequently was branded a hoax.

As months passed, hope naturally faded, and the *Atalanta* was declared lost with all hands.

What remained was to try to determine the causes. Speculations continued, but had nothing to support them. Storms were a logical theory, of course, and at that time of year it was almost a certainty that the vessel had run afoul of some of the Atlantic's nastier moods. Countering this theory, though, was the fact that several other ships were crossing at the same time; and they made it with nothing worse than delays. Besides, as a British naval vessel, the *Atalanta* was very well constructed and manned, and was considered to be more stable than the average merchantmen in heavy weather. Most significant of all against the storm theory, vessels so wrecked invariably left clues in the form of flotsam—planks, spars, boxes of one kind or another and miscellaneous debris. Often there also were survivors or the bodies of victims. There were none of these for the *Atalanta*, at least none definitely established as coming from her.

It was as though she had vanished into thin air.

It's of current interest to note that some observers now consider the *Atalanta* a victim of the eerie Bermuda Triangle.

For a vessel to disappear forever without a trace is quite a stunt. A more intricate performance is a ship that repeatedly vanishes and reappears—without a crew. Such is the story of the Hudson's Bay Company's steamer *Baychimo*. And this one has a mystery within a mystery.

The *Baychimo* saga began in 1921 when the 1,300-tonner went into service in far north waters. Her home port was Vancouver in British Columbia, but travels for her owners took her into hazardous polar seas where ice is a menace much of the year. For nearly a decade she performed her duties without mishap, something of a miracle, considering the conditions under which she operated.

But in October of 1931 the ice caught up with her. With a blizzard swirling about her, she found herself imprisoned in Arctic Ocean ice west of Point Barrow, Alaska, a fortune in furs stowed in her hold. As the ice gripped her more tightly, exerting enormous pressure, there was imminent danger of her steel body being crushed. Accordingly, her skipper and crew of sixteen left her for the time being, setting up a camp on the ice nearer shore but close enough to keep an eye on the steamer and her valuable cargo. If worse came to worse, they would wait until a spring thaw or the *Baychimo* was sunk by the ice, whichever came first.

A month later the *Baychimo* began her disappearing-reappearing act. It started when a storm was accompanied by somewhat warmer temperatures, warm enough to free her from the ice pack's grip. One morning a couple of days later when the storm slacked off, the men awakened to find their vessel gone, her moorings broken. Since the ship was not in sight and there was nothing they could do anyway, they went to a settlement at Point Barrow.

On the report of a hunter that the *Baychimo* now was about forty-five or fifty miles south of where they had last seen her, the crew enlisted Eskimo aid and traveled across the ice by dog sleds until they found her. It took the group several days to unload most of her cargo of furs, worth an estimated one million dollars. Then, on a return trip to finish the chore, they discovered that once again the steamer had taken off on her own.

Nearly six months passed without word of the errant vessel; then she was reported closed in ice near Herschel Island, which lies just off Canada's Yukon Territory in Mackenzie Bay, well east of Point Barrow. That summer, 1932, some prospectors boarded her. Whether or not they salvaged any remaining furs isn't mentioned, but they said the ship was in good shape. Despite recurring bouts with crushing ice, she was still very sound. She was not sighted again until the next year, this time by Eskimos. If she was enjoying her freedom, she seemed undecided about where it would take her. In the 1933 sighting she was passing Point Barrow.

In 1934 the wandering *Baychimo* was boarded by the inquisitive crew of a schooner. What they found was mostly junk. From 1934 on the derelict was sighted periodically every year, first here, then there. Another boarding party scrambled onto her deck in 1939. Apparently they had salvage in mind, but bad weather prevented any towing into port.

In his absorbing book *Invisible Horizons,* maritime chronicler Vincent Gaddis says that, so far as he could determine, the *Baychimo* was last observed in 1956 by Eskimos. At that time she was traveling northward in the Beaufort Sea, an arm of the Arctic Ocean shared by eastern Alaska and western Yukon Territory. As of then, she had been wending her merry way for a quarter of a century, all by herself.

Her now-you-see-me-now-you-don't performance was the *Baychimo*'s own peculiar puzzle. An enigma within that mystery was how she eluded a crushing death from ice for so long.

In all probability it finally caught up with her. On the other hand, maybe the old lady is just hiding in the far, far north.

Haunting Story of the "James B. Chester"

Maritime mysteries come in assorted types. There are vessels— aircraft, too—that vanish completely with all hands, for reasons beyond human comprehension. In another category are what we could call supernatural phenomena. In this department are phantom ships, canvas set and manned by spectral crews, which come and go like fog. (See *Ghost Ships*.) Then there's a type in which vessels remain behind but their people vanish mysteriously, without positive clues to their fate.

A classic example of the third category above is the brigantine *Mary Celeste*, whose curious, tragic history is detailed elsewhere in this book. So classic is the chronicle of the *Mary Celeste* that it has elbowed into the shadows a very similar case, that of the bark *James B. Chester.* Unhappily, the resultant obscurity has cost us a number of details. Happily, there's enough of the faded story to intrigue us.

The *James B. Chester* belonged to the same mid-1800s era, give or take a decade, as the *Mary Celeste*. Nowhere have I been able to find her dimensions, but she has been described as a three-master of respectable size. By coincidence (or was it something more sinister, a Bermuda Triangle sort of incident?), her mystery came to light in the same large oceanic expanse, with the Azores as its hub, as that of the *Mary Celeste*, and their stories are strangely parallel.

Act I

On a day in late February, 1855, the three-masted bark was sighted by a British vessel, the *Marathon*, approximately 600 miles southwest of the Azores, her sails set on a tack. From a distance there didn't appear to be anything peculiar about her, but as the *Marathon* drew closer she noticed something very odd indeed: No one was at the *James B. Chester*'s helm.

Ocean crossings under sail could consume many days, and even a passing encounter with another vessel out there on that lonely expanse was a pleasant, however brief, interruption of tedium. This was to be quite an interruption.

When the two ships drew close enough to each other, the bark was hailed by megaphone. It failed to draw any response. Strange. Even stranger, not one man could be seen anywhere on deck. Without a helmsman, she had silently continued on her course. Now the *Marathon*'s master was really interested. His attention was riveted by more than curiosity, however. Since the *James B. Chester* appeared to be a derelict, he had salvage rights in mind. Accordingly, he put a boarding party on her.

Act II

The bark was searched thoroughly. She was completely devoid of human life, on deck or below, and her cargo was undisturbed. But all had not been serene earlier. She exhibited dramatic evidence of frenzied activity. Her cabins were a mess: various articles scattered all over the place, drawers ransacked, tables and chairs overturned. Yet there were no signs of a fight, such as might have been provoked by a mutiny or capture by pirates. The boarding party had to conclude that she had been abandoned in great haste, but was at a loss to explain why. Deepening the mystery, the *James B. Chester* carried more than enough provisions and water, and all her ship's boats were secure in their davits, yet her compass and papers were gone.

Lacking bloody evidence, signs of real violence, and at least one or two corpses, mutiny was ruled out. So was piracy, for the same reasons plus the fact that her cargo was intact. Furthermore, pirates probably would have scuttled her to destroy any evidence. There had been no fire or explosion. The bark was perfectly sound—under way when sighted, remember. Considering her health and lack of damage topside, storms had to be eliminated as a reason. Besides,

9

sailors wouldn't be likely to leave their ship in a storm unless she were in imminent danger of going down, in which case they wouldn't waste precious time to create such disarray in the cabins.

A truly drastic situation is required to force seasoned mariners to abandon a vessel in midocean. With all the standard reasons ruled out, what had caused her crew to leave the *James B. Chester*—and in seeming terror and panic? Equally intriguing, *how* had they left her?

Act III

The derelict was brought into Liverpool as a salvage prize by some crewmen from the *Marathon.* There, for a while, she was a curiosity attraction at quayside. After that we lose sight of her.

As always in such cases, different theories were offered to explain the mysterious disappearance of the *James B. Chester*'s men, but no investigator had much to go on. Among the suggestions was a popular conjecture of the period: That the bark had been attacked by a sea serpent, or a huge octopus, or a giant squid. In those days mariners were convinced that such monsters could drag ships down to their doom.* It wasn't explained, though, how they could pluck men one by one, yet leave a vessel undamaged.

Nowadays there's a school of thought that blames such disappearances in the Bermuda Triangle on UFOs from outer space. Maybe that theory can be applied retroactively to the era of the *James B. Chester* and *Mary Celeste*.

What seems most likely is that the twin riddles of those vessels will remain as they have all along—unsolved.

---

*For centuries, seafarers' lore has been spiced with tales of sea serpents—great, snakelike creatures up to 100 feet long; and their existence is debated to this day. No known octopus grows large enough to menace even the smallest sailing ship. Besides, octopi are timid by nature. Any problems with giant squid are highly improbable, since the biggest known species, Architeuthis, thought to attain lengths up to sixty feet overall, usually frequents depths below 1,500 feet.

2

Ghost Ships

Do you believe in ghosts, phantoms, spooks, wraiths, poltergeists, specters, or apparitions by any name? If so, the sea offers ample support to your subscription.

Fact is, nagging evidence persists even if you do not hold with such beliefs. Even if we discount the well-known superstitious nature of old-time mariners and the possibility that some of their visions may have occurred while they were, shall we say, under the influence of the grape, evidence still exists that is difficult to ignore, impossible as it may be to explain. In this it's rather like the situation with UFOs.

In any case, whether you're a ready believer or a devout skeptic, you'll find the reports of eerie visions at sea fascinating. Relax and enjoy.

A Phantom Sea Battle

On February 4, 1754, some fishermen were drawing their nets across a broad expanse of bay in the eastern reaches of Long Island, New York. It was a scene of quiet industry as they emptied their seines, set them out again, and inched toward Block Island Sound.

Suddenly the netters' absorption in their work was shattered. Out of nowhere there appeared a frightening apparition. The fishermen felt the skin on the napes of their necks begin to crawl as they stared open-mouthed. Hard by them in full view loomed . . . But let an eyewitness describe it in his own words (from a contemporary newspaper account):

> . . . an appearance like three ships full rigged, with their sails spread, the largest of which had a Pendant at her Main-Top Mast Head. The persons who saw it were so near to them that they could plainly see the Men upon the Quarter-Deck, as also their Yards, Tops, Blocks, Rigging, and other appurtenances belonging to ships. And in a few minutes they seemed to engage each other; they could see the Smoak (sic) of their

11

guns, but hear no Report. In the Time of their engagement they put about several Times, and the larger ship hauled up her Courses; the whole thing was performed with the utmost dexterity, as tho' they were navigated by the most skilful *(sic)* Mariners. And thus it continued for a quarter of an hour and then by Degrees vanished away. As we look upon it to be a supernatural Production so the Consequences of it we shall not pretend to determine.

So was ghostly combat between phantom ships described by fishermen two centuries ago in a New York newspaper account datelined March 18, 1754.

Was it a fantasy? That doesn't seem likely, since the eerie affair was witnessed simultaneously by *several* solid citizens, men of unchallenged veracity and not given to heavy drinking while they worked. Was it mass hypnotism perhaps, such as has been suggested in instances of UFO sightings in our time? Not likely. The witnesses were soberly realistic men, engrossed in hard work to scratch out a living. They had neither inclination nor reason to be mass-mesmerized.

What then? Well, your guess is as good as anyone's, for the phenomenon never could be satisfactorily explained.

127 Years Later, a Sequel

In the summer of 1881 a New York *Sun* reporter was voyaging on a fishing schooner in the same general area where that spooky naval engagement had taken place more than a century earlier. As ghosts in general and phantom ships go, 127 years is not a long hiatus; but in light of that time gap it seems safe to say that the newsman was in no way influenced by an account of the previous spectacle. He may never have even heard about it.

On the night in question, a humid stifling one, the schooner dropped anchor in Gardiners Bay, a roomy body of water partly enclosed between two eastern Long Island peninsulas. Except for subtle swirls of an ebbing tide, the bay was a sheet of black glass. Not so much as a vague whisper of a breeze stirred the night air.

In the cabins below, sleep was impossible because of the clammy heat. The reporter joined the schooner's mate in stretching out on deck. Even there it was difficult to sleep, but after a while he managed to doze off. Next thing he knew, he was being awakened by the mate, shaking him roughly. Speechless, trembling in terror

12

and on the verge of panic, the mate pointed outward into the darkness. As soon as the reporter's eyes focused after sleep, they flew wide open in astonishment. In his newspaper account he wrote: "A big schooner was bearing right down on us at a ten-knot rate, and not a breath of wind in the bay. Then, just as it seemed the oncoming vessel would crash into us amidships, it vanished into thin air."

Presumably, any specters manning the phantom schooner enjoyed a laugh as they evaporated.

The reporter's hard-nosed realistic side tried to tell him that the apparition was some sort of phosphorescent phenomenon, or maybe something akin to the electrical "St. Elmo's fire" often seen dancing in ships' rigging. But he never quite convinced himself on that score. As for the frightened, voiceless mate, he accepted the vision at face value—a ghost ship.

The "Flying Dutchman"

What the *Mary Celeste* is to the mysterious vanishing of crews without a trace, the *Flying Dutchman* is to the chronicles of phantom ships.

The story of the *Flying Dutchman* is very old, dating back to the late 1500s. It's also very long, with chapters continuing into this century as recently as World War II, if we can judge by accounts.

Like many such tales coming down through time chiefly by word of mouth, several versions exist. They all share one thing in common: Sightings of a ghostly sailing ship named *Flying Dutchman*. Beyond that, versions vary somewhat according to the powers of observation of those who claim to have beheld her over the centuries, their personal interpretations, and the inevitable "embroidery" that accompanies telling and retelling. Offered here is a kind of composite version, based upon reports of sightings as late as the 1940s.

A Legend Is Spawned

A principal haunting ground of the *Flying Dutchman* is the waters off Africa's Cape of Good Hope. It's there that the weird tale began about 400 years ago.

Keystone of the original story was an unidentified Dutch ship that was lost without a trace while sailing from the Atlantic Ocean into the Indian Ocean off the Cape of Good Hope. Quite probably she was the victim of a storm. Waters become very turbulent in that region where two oceanic masses collide. In any event, she seems to have inspired the name of the phantom that was to haunt the seas for the next four centuries.

The first mentioned sighting of the ghost ship occurred an unspecified number of years later when two Dutch merchant vessels were rounding the Cape of Good Hope. According to their account, the specter abruptly appeared out of nowhere; and to make the vision doubly impressive, the ghostly vessel looked to be battling her way through a fierce storm, even though the weather was fair and the seas relatively calm in that area at the time. No name was visible on the phantom's stern. Apparently this was when the appellation *Flying Dutchman* was bestowed, based on the fact that the aforementioned vanished ship had flown the Dutch flag. The "Flying" probably was inspired by the rolling and tossing of the ghost fighting a gale. In time the phantom's skipper also acquired a name, although no one knows when or how. There are different versions of this too, but the most common in narratives is Vanderdecken.

The two Holland merchantmen fled the area, convinced that they had seen the ghost of the vanished Dutch ship. But one of several speculations as to the spook's identity was proposed in an obscure French book about the sea published in 1832.

It is that volume's contention that the *Flying Dutchman* was beating her way around the Cape of Good Hope when she ran into vicious weather, with violent winds and battering waves. Winds tore at her rigging, reducing her sails to tatters and carrying away spars. One by one her masts toppled (the *Flying Dutchman* has been variously described as a bark, a schooner and a brig). But none of this destruction and impending doom bothered her master; and when crew and passengers pleaded with him to try to find haven from the storm he refused, punctuating his tirade by killing their spokesman and tossing him overboard.

Then a visionary figure descended from heaven and materialized in human form on the quarterdeck, legend has it, and remonstrated with the captain. Crew and passengers were paralyzed by awe and fear of The Vision, which spoke with obvious authority. But the stubborn captain merely laughed, answered rudely, and finally threatened to shoot the heavenly visitor. He did, in fact, draw a pistol and fire, but The Vision was unharmed.

As a penalty for this hostility The Vision declared that henceforth the evil skipper and his ship were accursed, condemned to sail the seas forever and ever without rest or port. The passengers, presumably, were excluded from this condemnation, although we hear nothing more about them. Also exempt was the crew, all except a cabin boy. For some reason, possibly because he was a disciple of the demonic captain, the cabin boy also was sentenced to remain aboard for eternity. In a final pronouncement, The Vision declared that the captain now had become an evil spirit of the sea, adding (very unreasonably and unfairly) that henceforth misfortune would be the lot of all ships and all sailors that might so much as sight the accursed vessel.

In Music, Prose and Superstition

Composers, poets and writers have gotten a lot of mileage from the *Flying Dutchman* legend over the years. It's believed to have been the inspiration for Richard Wagner's *Der Fliegende Hollander*, which freely translates as *The Flying Dutchman*. That anglicized version also was the title of a drama penned by Edward Fitzball in 1828. Earlier there was an anonymously authored piece of fiction in a Scottish magazine that was titled *Vanderdecken's Message Home*.

Sir Walter Scott wrote a version of the story. So did English naval commander and sea-life novelist Frederick Maryat (1792-1848), who wrote under the name Captain Marryat (with a double "r"). His interpretation was entitled *The Phantom Ship.* American-born British novelist William Clark Russell (1844-1911), joined the club in 1888 with *The Death Ship.* Samuel Taylor Coleridge drew upon the *Flying Dutchman* for his immortal *The Ancient Mariner*, as did John Greenleaf Whittier for *The Dead Ship of Harpswell*.

Many writers, notably those of the nineteenth century, have collaborated to give the *Flying Dutchman* legend durability. From Europe has come yet another version.

In this one the Flying Dutchman began as a person, a thirteenth century nobleman, who murdered his brand-new bride and his brother when he discovered them together in what he thought was adultery—on his wedding night, no less. Later, so the story goes, he learned that he had been wrong.

Today the nobleman's act would be termed justifiable homicide and he probably would go scot-free—to write a book that would become an overnight best seller, be made into a movie, and earn him a million dollars. But the thirteenth century aftermath was quite different.

The nobleman was condemned to forever wander northward. This roving eventually brought him to a beach near Haarlem in the Netherlands. There he met a stranger who had been awaiting him with a boat. Beyond at anchor lay a ghostly bark. On command the wretched nobleman got into the small boat and, now accompanied by two specters, a "good spirit" and an "evil sprit," rowed out to the phantom vessel. He assumed command and put to sea at once, all canvas set.

And so the condemned man, now a skipper, was destined to sail the bark forever (to date he has logged 700 years), while the spectral good spirit and evil spirit tossed dice to see which would eventually claim his soul. Always they sail to the north, which would seem to pose serious geographic limitations but which hasn't hurt the legend. This *Flying Dutchman* sails not off the Cape of Good Hope, but in the North Sea. The phantom is readily recognizable, 'tis said. Her hull is a drab gray and her sails are of various colors. No one is ever at her helm, nor can any other crewmen be seen as she continues her restless northward course.

A thought occurs that possibly this *Flying Dutchman* negotiates the ice fields of the North Pole to continue her voyage on the other side of the world until she eventually reaches the Cape of Good Hope, then goes on north again. Or maybe there are two *Flying Dutchmen*, one to haunt the Northern Hemisphere, the other to scare sailors in the Southern.

In yet another version of the origin of the *Flying Dutchman*, the ghost is believed to be that of a Dutch East India merchantman, *Libera Nos*, commanded by a Captain Bernard Fokke. The fate of the *Libera Nos* isn't recorded, but her wraith has been reported in various areas down through the centuries. Her master and crew are all skeletons, say reports. Could this be an impostor, cashing in on the real *Flying Dutchman*'s publicity?

Offshoots

As if the enduring legend itself were not enough, it has sired an unending series of superstitions, all evil. Here are a few samples:

1. The *Flying Dutchman* leads ships to destruction on rocks and reefs, or causes them to sail into strange waters where they encounter calamity in one form or another.

2. The phantom's evil skipper can cause other ships' water supplies to become unfit to drink, their wine to turn to vinegar, and their food supplies to rot.

3. The worst of all misfortunes occurs if the *Flying Dutchman* is allowed to draw alongside. Sometimes, it has been said in hushed tones in fo'c'sles, the phantom ship can be seen to launch one of her boats, variously described as being empty or rowed by skeletons. If this boat is allowed to come alongside, disaster is certain to follow.

And there's an insidious note: It has been whispered about that Captain Vanderdecken can alter the *Flying Dutchman*'s appearance at will so that victims can be caught unawares.

I put the foregoing in the present tense because—who knows—

the *Flying Dutchman* may still be with us. She's still young for a ghost.

Sightings

There have been many, and not all confined to the Cape of Good Hope region.

In July 1881, the British war vessel *Bacchante*, on cruise with her squadron, duly recorded seeing the *Flying Dutchman* in her log. In this meeting the spook was a brig. She appeared first as a large, red glow; but then, as she drew within a few hundred yards, her masts, spars and sails became plainly visible. The *Bacchante*'s bow lookout witnessed the apparition, as did an officer on watch on the bridge, and some crewmen—a dozen or more in all. Signals from two of her sister ships asked *Bacchante* if she had seen an eerie red glow—and what was it? Then, as suddenly as it had appeared, the apparition evaporated. All this took place on a morning in fair weather, on a calm sea.

The *Bacchante*'s logbook went on to say that it couldn't be determined whether the spectral bark was the famous *Flying Dutchman* or another ghost ship haunting that region. But the aftermath was in harmony with the former's legend. Later that day, according to the British ship's log records, a seaman who was the first to report the apparition fell from the foremast and was killed instantly. When the squadron reached port, the *Bacchante*'s commanding officer was stricken with a strange illness and died.

Ubiquitous Phantom

Spring, 1866, saw an Australian sailing ship, the *General Grant*, long out of her home port of Melbourne on a voyage to London. For several nights she had been followed by a ghostly-looking vessel thought to be the *Flying Dutchman*. On May 13th she found herself in a weird situation: In an area normally swept by trade winds, they experienced not so much as a gentle zephyr; yet the ship continued to move as though propelled by an unknown force. What was worse, this force drove her onward until she was literally jammed into a huge seaside cavern in the high headlands of Disappointment Island (note that name) in the Pacific, and there she sank, with a heavy loss of life.

Her safe was carrying about a million dollars' worth of gold dust when she went down in the cavern. Word of this disappearance spread, and the *General Grant* became a magnet for get-rich-quick salvagers. But the *Flying Dutchman*'s legendary curse was transferable, apparently, because none of the retrieving operations was successful.

17

And the curse continued its blanket coverage of other vessels. In 1870 the schooner *Daphne* put out from New Zealand with one of the very few survivors of the *General Grant* disaster on board. Their objective: getting that gold in the wreck's safe. On arrival at Disappointment Island, a smallboat containing the survivor, the *Daphne*'s skipper, and three seamen rowed toward the giant water-side cave to pinpoint the gold ship's location on bottom. They were observed entering the cavern, but they never emerged and were never found. The *Daphne*, it goes without saying, didn't salvage the gold.

When she returned home to New Zealand, the vanished small-boat and men naturally had to be reported. The crew then testified that while lying at anchor off Disappointment Island, a phantom ship, supposedly the *Flying Dutchman*, was seen bearing down on them with all sails set. Just as a collision seemed imminent, the ghost vessel altered course to cross the *Daphne*'s bow, and prompt-ly disappeared.

In a follow-up to the *Daphne* story, eighty-eight years later in 1958, the vessel of an English salvage company also tried unsuc-cessfully to get at the seemingly forbidden gold in the *General Grant* . . . and underwent a similar encounter with a phantom sailing ship.

Contemporary newspapers reported sightings of the *Flying Dutchman* through the last decade of the 1800s and on into the first two decades of the 1900s.

Whose Curse? Gold's or the Ghost Ship's?

In the late 1890s a steamer named *Hannah Regan* was traversing the South Pacific with a cargo that included approximately $1,000,000 in gold bullion, locked in a safe. In the vicinity of Okinawa, an island that was to become infamous in World War II half a century later, she ran into extremely heavy weather that cost her a propeller and other severe damage. Attempts were made to effect emergency repairs so she could at least limp out of that place, but something went awry with the missing propeller's shaft and the *Hannah Regan* sank. It was from the captain's personal log, found with his body (along with the corpses of the first officer and four other men, drifting in a lifeboat several weeks later), that the ill-starred steamer's fate was learned.

A million dollars in gold is a fat prize. It drew a salvage project on a seagoing tug out of San Francisco, which located the wreck and commenced salvage attempts. In his report of their operations the tug's skipper mentioned the following incident.

On a clear, calm night, while the salvage crew rested between days of hard labor, the captain took a turn about the deck before hitting the sack himself. As he watched the buoys marking the wreck of the *Hannah Regan* bob in the moonlit water, he suddenly became aware of what he first thought was a large, moving shadow, then about a half-mile away.

He watched it approach, and his curiosity gave way to amazement as the shadow gradually assumed the form of a sailing vessel. She was plainly discernible, he said, and of a type that hadn't graced the seas in two centuries or more. On she came, headed for the tug. She moved as though being driven by a strong wind, yet she carried no canvas; and although the Pacific was flat calm in the tug's area, the mysterious ship rolled violently, as though battling a stormy sea. Waves crashed over her as she approached, and she looked to be sinking by the stern. When it appeared that this mad vessel was about to run down his tug, the captain's instinct was to shout a warning. At that instant, however, the realization struck him that this was an apparition.

The phantom's stern appeared to be submerging as mountainous waves dumped tons of water on her. Rooted to the spot, the captain watched as the specter drew alongside. At this point, he related, he thought he was losing his marbles, because while he could see completely through this ship, details of her rigging and deck were clearly defined. Two of her smallboats were being dragged from their davits as she glided by.

The tug captain's account had an unusual ending for a *Flying Dutchman* sighting. As he continued to watch in open-mouthed astonishment, the phantom ship disappeared beneath the waves like a diving submarine.

Salvage operations were resumed the next day. It was then that a *Flying Dutchman* curse asserted itself. A diver had descended with an explosive charge in an effort to blast open the jammed door of the safe containing the *Hannah Regan*'s gold bullion. When the diver failed to surface and there was no explosion, a second diver was dispatched to investigate. He surfaced to report that the first man's air hose was fouled in the wreck and was severed. The second diver redescended to see if he could help. Both were lost.

To cap the double tragedy, the salvage tug never did recover the *Hannah Regan*'s gold.

More in Our Time

The most recent reports of *Flying Dutchman* sightings this writer has come across so far are of World War II vintage—if it was indeed the famous phantom. These sightings reportedly were made by crews of Nazi submarines at various locations east of the Suez

Canal. That would be quite far north for our first *Flying Dutchman*, whose favored haunt is the Cape of Good Hope. But then, this phantom does seem to be far-ranging. Or maybe it was our northern edition of the spook ship, sailing south for a change after all these years.

A Ghost Ship Comes Ashore

Among all the tales about phantom vessels I have heard and read, in only one did the spectral ship approach shore. This vessel even put a man on the beach. And it all happened in broad daylight. The observer was none other than prolific author Percival Christopher Wren (1885-1941), whose novel *Beau Geste* was a bestseller in its day and ultimately became a motion picture. Originally his first-person account was published in a British magazine. It has since been retold in books dealing with psychic and supernatural experiences. Vincent Gaddis included it in his *Invisible Horizons*.

When he was young, not yet famous as a novelist, Wren was wandering along a beach on England's eastern coast one summer morning. Suddenly he became aware of the presence of a strange vessel not far off the beach. She looked as real as any ship he had ever seen, he commented. She had a high bow, with a figurehead suggestive of a dragon. Her stern also was high. The single mast was rather short and supported a dirty, tattered mains'l which bore an undecipherable symbol. Outside along each gunwale he could see a row of shields, such as were used by Vikings in hand-to-hand combat. They too were the worse for time and weather.

Seated athwartships were four rows of oarsmen. Standing in the stern was another figure, apparently the helmsman. Wren could make out still other men in various postures on deck. He said he could plainly hear their shouts as the strange craft was maneuvered toward shore.

The vessel eased into shallow water near the tide line, whereupon a few of the men picked up one of their shipmates, who had been sitting slumped against the mast, and unceremoniously threw him overboard. As the craft moved back out into deeper water, the marooned sailor staggered to his feet and collapsed on the sand.

All this was observed by Wren, who now took a closer look at the man on the beach. His face was weather-darkened, the author said, contrasting sharply with a white beard and mustache. An odd kind of metal hat, not identifiable, crowned a mane of white hair. So real was this stranger's appearance that Wren rushed toward him to raise his head, lest he drown at the water's edge. But as abruptly as the vessel had appeared, the man "evaporated." And when Wren looked to sea, the little ship also had vanished.

The novelist stated that he definitely saw and heard this eerie tableau, and not in a daydream or hallucination. As it unfolded it couldn't have looked more real to him.

Legend of the "Half Moon"

In 1609 English navigator-explorer-adventurer Henry Hudson undertook a voyage to the New World for the Dutch East India Company. His ship was the *Halve Maen*, better known by its anglicized name, *Half Moon*. On this expedition Hudson discovered the New York State river that was named for him and ventured up it as far as what is now the capital city of Albany.

I was unable to track down its source and age, but an obscure legend has it that the ghost of the *Half Moon* sometimes can be seen at night, visible in a pale glow as she makes her way along the broad bosom of the Hudson.

Perhaps that accounts for a saying in the great river's valley: When thunder rumbles, it's Henry Hudson and his men enjoying a game of lawn bowls.

Of Later Vintage

The United States had been engaged in World War II not quite a year that November day when the destroyer *Kennison* glided cautiously toward San Francisco through a pea soup fog. As she approached the coast her radar beamed signals ahead, searching for the Farallon Islands, about thirty miles off the Golden Gate. At intervals the throaty blast of her foghorn warned other ships of her presence. Three lookouts were on watch: One forward, another on the after gun deck, and a third on the destroyer's fantail.

All at once the sailor on forward watch heard an alien sound off the *Kennison*'s port quarter. As it grew closer and louder he interpreted it as that made by a ship's bow knifing briskly through the sea, but he could see nothing for the fog. His binaural sense—that is, his two ears acting as a sound-locator—told him the vessel would cross the destroyer's wake close astern. While he was pondering this, the fantail lookout's excited voice came over the intercom, calling to the bridge to look aft. Then the lookout on the after gun deck cut in, announcing that he also saw something aft.

In an exchange that bounced back and forth between the *Kenni-*

son's bridge and the lookouts it developed that a two-masted sailing ship of long-ago design had loomed out of the fog momentarily, crossing astern of the destroyer and barely clearing her. She was a beat-up vessel, said the watch who had got the best look at her. He described her as being under full sail, although her canvas was tattered and her rigging decrepit. He could see her wake as she passed, he declared, and he heard the creak of her timbers. No crew was evident on deck, not even a man at her wheel.

A lookout had seen and heard the two-master, but when she passed the *Kennison* no blips registered on the destroyer's radar.

Ghost Ship with a Sailing Schedule

Great joy precedes tragedy in the romantic but fatal saga of the English schooner *Lady Luvibund*.

On the thirteenth day of February in the year 1748 she hoisted her sails for a routine cargo voyage to Portugal. That is, it was a routine run for the schooner, but for her master, Simon Reed, it was a red-letter trip. He had chosen this voyage for his honeymoon. And as the *Lady Luvibund* set her course for the Iberian Peninsula, his radiant bride and he were joined by a few invited guests for a shipboard party in his cabin to mark the occasion.

Already there was bad news topside. What Captain Reed didn't know was that he had taken away the love of his life from his first mate, John Rivers, and in his ignorance had innocently invited Rivers to be his best man at the wedding. Now the first mate was becoming mindless with hate, jealousy, frustration, and mounting fury as he listened to the sounds of celebration coming from Captain Reed's quarters. Soon his emotions pushed him beyond all reason. He would have revenge, whatever the cost. (A threat in this vein was mentioned by Rivers' mother in her testimony at an official inquiry.)

Masking his insane rage, the first mate casually strolled over to the helmsman and volunteered to relieve him for a spell, a recess which that worthy welcomed. Now alone on deck and with sounds of revelry still in his ears, Rivers swung the schooner's bowsprit toward shore. With sails billowing, the *Lady Luvibund* moved briskly toward a date with destiny and soon was driven with a jolting crash into a sand bar. The impact must have been great, for the unfortunate schooner sank very quickly. The wedding party was drowned in the midst of celebration. No one aboard escaped death. At the inquiry which followed, fishermen testified to witnessing the *Lady Luvibund*'s suicide.

At this point readers with a superstition about the number 13 can say, "Well, what do you expect? She sailed on the 13th." They'll have occasion to point to that number again as we continue.

On February 13th, fifty years to the day after the *Lady Luvibund*'s demise among the shoals, an irate Captain James Westlake, master of the coastal vessel *Edenbridge*, heatedly petitioned authorities to apprehend and punish the officers of a crazily operated schooner that had narrowly missed his vessel. The near-collision had occurred in the vicinity of the *Lady Luvibund*'s sinking half a century earlier. Subsequent investigation revealed that the offending vessel was an apparition. Fishermen testified they saw a ghost schooner collide with the sand bars and break apart quickly.

The incident was forgotten until another fifty years went by. And then, on—you guessed it—February 13, 1848, the crew of an American vessel, along with several witnesses in local craft, saw a schooner run aground, break up and sink in the same place. Moments before, they said, they heard a woman's laughter and party sounds float across the water from the oncoming schooner. The spectacle must have looked very real to the observers, because they hurried to the scene to search for survivors. There were none, of course, nor any signs of a wreck or debris among the shoals.

Records show that the spectral *Lady Luvibund* (it's assumed that the ghost is hers) has been putting on the same fatal performance on February 13th at fifty-year intervals ever since. If those "bookings" continue, the next showing is scheduled for 1998. Look for news of a wreck that year.

A Phantom Etched in Fire

While this book was being written I met a Boston College professor of history and his wife who had recently visited Nova Scotia. Since he also writes books, we fell into a discussion of the trials and tribulations of authorship, and he inquired about the book you are now reading. When I described its contents he mentioned a legend he had heard in Nova Scotia. It so happens that I had heard it too, while in Canada's Maritimes several years before. The story has stayed very much alive—nourished, apparently, by the almost regular repetition of a spectacular apparition.

The scene of this ghost ship's appearance is the area where Northumberland Strait, sweeping between Prince Edward Island and Nova Scotia, flows into the Gulf of St. Lawrence. It's said that sightings have been reported by coastal hamlets along the strait, notably Merigomish. It may be appropriate that not far from Merigomish is a place called Malignant Cove.

At any rate, this particular spook favors timing its arrival close to that changeover of seasons known as the autumnal equinox; and always after sunset. She appears out of nowhere as a three-masted ship under full canvas, usually on an easterly or northeasterly course. Lights are seen in her rigging and on deck. If the moon is bright enough, shoreside observers have declared, speckles of light glint on her copper bottom when she heels on a tack. In fog, a common condition in those parts, weird phosphorescence outlines her.

The ghost ship has never been identified by name, but it puts on quite a show. The phantom glides on briskly to a certain point (near Malignant Cove, perhaps?) where it abruptly lurches, as though hitting or sideswiping a reef. In a pyrotechnical grand finale, flames suddenly erupt and envelop the vessel, and dim figures of her crew can be made out leaping into the sea. When fire has consumed the ship and her sails, the flames die down, and her gutted remains disappear beneath the waters of Northumberland Strait.

3

The Hanging of Albert Hicks

In July of 1976 hundreds of thousands of spectators converged on New York Harbor by every conceivable conveyance, shank's mare to private aircraft, to witness Operation Tall Ships, a magnificent display of sailing vessels amassed as a special salute to the United States' 200th birthday.

In a July 116 years earlier, that general area also drew throngs of rubbernecks from far and wide—not as many as for the 1976 spectacular, but a lot for an era of horses and buggies. On this occasion the magnet also was a maritime attraction, the steamship *Great Eastern*, considered a marvel in ocean transportation in her day.

You could say that on July 13th, 1860, the area was offering a double feature. On that day a man named Albert Hicks was scheduled to be executed publicly, and it was open house for everyone who wanted to watch. Hicks's exit was to be by hanging, from a gallows on Gibbet Island* in New York Harbor. He had been convicted of a vicious multiple murder at sea.

In company of a federal marshal and guards, Hicks w placed aboard a steamer, *Red Jacket*, for transfer to Gibbet I id. Accounts had it that the steamer's rails were lined with p op going to the hanging, a turnout which may or may not have pleased the condemned. At this point in the proceedings the authorities aboard the *Red Jacket* thought it might be a nice diversion to take a look at the *Great Eastern* while they were at it. This unusual excursion meant a two-hour postponement of the hanging, which annoyed the thousands who had turned out in holiday atmosphere—an incongruity of public executions long ago—to see him dance on air. The delay didn't irk Albert, who was in no sweat to be ushered into eternity.

After her passengers had had a look at the *Great Eastern,* the *Red Jacket* threaded her way downriver through a converging fleet of private and commercial vessels, some from as far away as Connec-

---

*An early name for a place that was to become much better known as Ellis Island. The early name stemmed from its use as a site for execution of pirates, starting along about 1765.

ticut, that would take up positions as floating grandstands to watch a murderer pay for his crime. To add another macabre note, the vessel on which the brutal killings took place, the oyster sloop *E. A. Johnson*, freshly painted and laden with spectators, also rode her anchor line in view of the scaffold. Still more people crowded Gibbet Island's shores. It was estimated that upwards of 10,000 people turned out specifically to watch this "necktie party."

Once he arrived at the gallows, there were no more reprieves for Albert Hicks. In company of a clergyman and authorities he stood docilely while the hangman adjusted a hempen noose around his neck. An awed hush came over the crowd, electric with expectancy. With a nod from an official, the gibbet's trapdoor dropped open, and Albert Hicks plunged into eternity, his body quivering and twitching at the end of its rope. His victims had been avenged in what may have been the last hanging for piracy, or piratical murder, in the United States, a distinction of dubious value to the former Rhode Islander.

Where Albert Went Wrong

The offense for which Hicks was executed was one of blood-spattered violence—cold, brutal, calculated homicide. But the story began as a mystery.

On the previous March 15th, the *E. A. Johnson* departed a New York Harbor wharf for a northern New Jersey port, a relatively short haul. In command was Captain George Burr, a Long Islander. Aboard were two Long Island brothers named Watts, and a mate, calling himself Nicholas Clock, who had signed on at the last minute.

The oyster sloop's mystery did not begin until March 21st, well out in the New York Bight, a triangle of ocean between Long Island and the New Jersey seaboard. Early in the morning of that day she was found adrift by two vessels, seemingly derelict. Her sails were down, but unfurled and draped over her rails. Her bowsprit had been snapped off and it trailed in the water alongside. (Later it was learned that the bowsprit was broken off during a collision with a schooner in the dark, just a few hours before finding the oyster sloop adrift. Only one man was seen on her at that time, but apparently the two vessels didn't pause to compare notes.)

She was boarded for investigation by the captains of the two discovering vessels, since there had been no replies to their hails and no one was in sight on the obviously distressed sloop. The scene that assaulted their eyes was enough to tilt even strong stomachs. Pools of blood splotched her deck like big polka dots. Gore also was splashed about, and trails of it led to the rails. Below deck was another scene from a slaughterhouse. Blood was all over the place, spattered high and wide. In a cabin lay a red-smeared hammer, matted with human hair. No one was aboard, not even a corpse. There could be no doubt that her people had met with foul play; but under the circumstances the only hope of finding even one *corpus delecti* lay with cooperation from ocean currents.

The *E. A. Johnson* was towed into New York, where police and a coroner went over her with a fine-toothed comb. From the companionway ladder to the cabin below, every article and every surface examined seemed to be spattered or smeared with blood, even galley utensils. There was also evidence that a hatchet or knife had been used in the carnage. And everything was in violent disarray; the cabin was a mess. Further checking failed to turn up the sloop's papers, and there were indications that robbery may have prompted the ransacking. The sloop's boat, a small yawl, customarily hung from davits in her stern, but now was missing. Also missing, of course, were the victims of this mayhem. "Mystery Tragedy!" screamed newspaper headlines.

It was not too difficult for police to determine who had been aboard the *E. A. Johnson* when she set out on that terrible trip March 15th. They hit a snag, however: No one seemed to know anything about the mate who called himself Nicholas Clock. In the investigation he emerged as a shadowy figure who merited closer scrutiny. But where to find him? Had he been a victim in the multiple murder, or had he simply disappeared, very much alive, from the sloop? The more they probed, the more police became convinced that the slaughter was the work of an individual, the man glimpsed at the sloop's helm when she collided with the schooner during the predawn hours of March 21st. Moreover, they felt sure the villain was the mysterious Nicholas Clock, and theorized that he had deliberately collided with the schooner in a desperate attempt to sink the *E. A. Johnson* to conceal evidence.

Their theory gained substance from reports of a stranger answering the description of the missing mate coming ashore in a yawl the morning the murder vessel was discovered. He was the only survivor, he said, of a collision with a schooner that killed his skipper and sank his vessel before his shipmates could escape. From Staten Island the man who called himself Nicholas Clock was traced to lower Manhattan. Tracing him wasn't too hard. Along his way the stranger had made stops during which he displayed much more money than an ordinary seaman would be carrying. Several people remembered him. When he was recognized as a member of the *E. A. Johnson*'s crew by a man in the oyster business in lower Manhattan, police started to close in.

"Nicholas Clock," as you've undoubtedly guessed, was Albert Hicks.

It became more and more evident that Hicks wasn't totally sane. In his simple lodgings in Manhattan he had left personal possessions of Captain Burr and the Watts brothers. Hicks wasn't there at the time, but police readily tracked him down. Either stupidly or brazenly, he chose to return to his real home in Rhode Island.

Damned by Evidence

Confronted with the personal effects of Captain Burr and the Watts boys, along with the bloody hammer found on the sloop, Hicks confessed—but with an earnest explanation. Yes, he admitted, he killed the captain and the other two (he mentioned in passing that the skipper was a tough man to dispatch, which accounted for the cabin looking like a disaster area). But the killings were not really his idea. He had been prompted by no less than Satan himself. With Beelzebub's voice commanding him and furnishing how-to instructions, he methodically stalked Captain Burr and his shipmates with a hammer and hatchet, beating out their brains or hacking them to death, one by one. Then he threw the battered, messy cadavers overboard. He could give no tangible motive for the multiple slaying. Obviously, he was suffering from a mad delusion, caused by a persecution complex, perhaps, or paranoia, or maybe Hicks just didn't like his companions. As we've seen, there was some thievery too, which placed the case in the category of piracy; but theft appeared to have been more an afterthought than a motive.

The most logical explanation, of course, is that Hicks was violently insane. That may also explain why he didn't go to any lengths to hide the stolen personal possessions and flee to parts unknown, instead of lingering at his home, the first place police would look. In any event, he was convinced that the Prince of Darkness directed the massacre, and so told the authorities. They didn't buy that story. Apparently they didn't think that he should be treated as insane either.

Albert Always Kept Busy

But the route to the gallows on Gibbet Island wasn't complete yet. It developed that Albert Hicks had been similarly engaged on another occasion. This came to light while he was in custody for the *E. A. Johnson* blood bath, and it suggested strongly that homicide was a kind of hobby with him.

Sixteen years earlier a vessel named *Saladin* went aground near Halifax, Nova Scotia. Investigators discovered that her captain, mate, and two seamen were missing. The crew told questioners that the two officers died at sea (not completely a big fib, as it turned out) and that the two missing sailors fell overboard accidentally. Certain details of their story didn't ring true. Authorities became suspicious and held the survivors for trial. When subsequent findings pointed to a quadruple killing on the *Saladin*, some of the crewmen were convicted and executed for piratical murder. Another seaman—guess who—managed (in a way not made clear) to extricate himself from this very serious situation and went free. The *Saladin* case was closed.

A decade and a half later it was reopened by a confession. Held for piracy and murder in New York and realizing his life was lost anyway, Albert Hicks decided to let police in on what really happened to the four missing *Saladin* men. He and Satan were partners in that operation too, he admitted. With the devil leaning over his shoulder and giving him do-it-yourself butchering directions, Hicks killed the captain, mate, and two sailors and consigned their bloody remains to the Atlantic.

How this quadruple homicide escaped notice by others in the crew, if indeed it did, and whether or not anyone else was involved, isn't clear. But Hicks capped his confession with an announcement that chilled his listeners: Innocent men had been hanged for crimes he committed on the *Saladin*.

4

Legend of the Money Ship

This is a tale which, for fullest flavor, should be told in the yellow glow of kerosene lamp light in a weatherbeaten shanty beside the ocean, with the ageless song of the surf as background music and the wind rattling windows in ghostly raps as it moans around the eaves. It's a yarn that could have inspired Robert Louis Stevenson's *Treasure Island*, but, unlike that classic, is not fiction. The events it relates took place long, long ago, and the account has been handed down by generations of seafaring folk on Long Island.

Opening Scene
Our drama begins in the year 1825. The precise locale of its first scene has been obscured by the mists of time, but unquestionably it was a stretch of lonely beach beside a bay on eastern Long Island.

One summer day in that long ago, two young boys romped among the sand dunes as their fisherman father made emergency repairs to his gear on the beach. Finally wearying of chasing each other, the brothers started back toward the bay's shore, threading their way among little sand hills. Then, as they rounded the base of a dune and entered a deep hollow, they saw something that froze them in their tracks. In horror and disbelief they stared at a human skull and bleached bones, partly buried in the sand.

When their initial shock had passed, the boys raced back to their father and excitedly told him about their macabre find. The bayman accompanied them to the hollow and saw for himself. Brushing sand away, he examined the bones in silence. With a grunt, he tossed the skull aside and rose. Taking his two sons in hand, he led them away from that place. The skeleton, he told them, was the remains of a long-dead pirate. He didn't elaborate, but merely said that it wasn't good for little boys to know about such things.

If the father thought they would forget, he didn't reckon with his older son. Despite his parent's vague warning, the boy couldn't shut the matter out of his mind. His curiosity was piqued and kept nagging him. As time went on he quietly made inquiries away from home, asking here and there if anyone knew something—

anything—about that skeleton in the deep hollow among the dunes. No answers came.

Years passed and the boy entered early adolescence, tenaciously persisting in his inquiries. Along the way he patiently collected splinters of information until he was able to assemble a fragment of the story. It wasn't much to go on, but it told about a small group of buccaneers whose mortal remains had been uncovered by the wind in a dune hollow which lay somewhere to the west of a forbidding structure known as The Old House. Then came his first substantial lead. He learned that there was an aged seafarer, Captain Terry by name, who lived alone in a shack on a remote beach, and that this old-timer could tell him what he wanted to know.

His excitement mounting, the lad could hardly wait to visit the ancient mariner and ply him with questions. But two things gave him pause: It was a walk of several miles to Captain Terry's shack; and the way led through a dark, wooded area that everyone knew was haunted by the ghost of a slave who had been murdered there by his cruel master. The lad would have to pass through this spooky place after dark on his return home. But finally his determination overruled his fear of any ghost, and he undertook the long hike to Captain Terry's shack.

A Mysterious Stranger

In a voice cracking with age Captain Terry welcomed his young visitor. "Out here I don't see many folks," he said. He listened intently as the boy told the purpose of his visit. Then he poured himself a stiff snort of West Indies rum—"For my rheumatiz," he explained—and launched this narrative.

Many years before, toward the close of the 1700s, a strange ship hove-to off Montauk Point on Long Island's easternmost tip and put a man ashore. This done, the vessel stood out to sea and tacked to the west'ard.

The man on the beach was tall and brawny. His ugly face added to an aura of evil about him. He carried no sea bag, nothing, and started walking along the surf at a brisk pace. When he had covered about half the distance to Napeague Beach* he stopped near a large rock and made some observations. Satisfied with these, he signaled his ship, which had reappeared and now was lying-to not far off shore. The vessel answered by clewing up her fores'l, whereupon the stranger resumed his brisk walk, this time in the direction of Amagansett.

*This was the general area in which a Nazi submarine landed a small group of saboteurs during World War II.

31

It was dusk when the sailor reached that hamlet. Here he met with something less than hospitality. Little villages were isolated in those days, and their residents developed an inherent distrust of strangers, especially evil-looking ones. He was refused lodging for the night. Perhaps the people of Amagansett sensed he was a pirate and a man of violence. Muttering darkly, he strode out of the hamlet, and where he spent the night was never learned. However, word of his presence spread throughout the area. In those times, few things excited talk more than the appearance and disappearance of a stranger. The villagers of Amagansett, Southampton and East Hampton speculated for days on the sailor's identity and mission.

After leaving the Hamptons, Captain Terry said, the burly mariner was seen making careful observations along the coast below Ketchabonack. After that he vanished again, later reappearing miles away on a lonely stretch of road between Forge River and a place called The Mills. Along the way he made an inquiry of a Mr. Payne, an old soldier of the American Revolution. When the vicious-looking stranger had left, Mr. Payne's family asked him, "Who was that man?" To which he replied cryptically, "That I cannot tell. But one thing I know. Whoever he is, he has been in a human slaughter." The elderly war veteran said no more.

That Forbidding House

When the stranger again walked into Captain Terry's narrative it was in an unnamed hamlet beside a bay on Long Island's southern coast. There he was known to certain residents who thought he was serving as a ships' pilot during his long absence. To this tiny community he had returned to rejoin a wife and daughter who had a local reputation for unsociability and general unwholesomeness. This poisonous pair dwelled apart on the beach in a bleak, frowning structure which from colonial days had been called The Old House.

Some time later the big seafaring man returned to the oceanfront, where he found his ship biding her time as she awaited a signal from shore. It was obviously a prearranged rendezvous. The sailor's message was that she should stand by until nightfall.

As the afternoon advanced, a sou'east squall whistled in swiftly. Amid sheets of rain, whitecaps did a wild dance and wind blew spume from crest to crest. It became a scene of great menace, but the mysterious ship maintained her position as best she could. After dark, when the storm had abated somewhat, a large fire was ignited on the beach. It was the awaited signal, telling the vessel's crew when and where to come ashore.

Despite the heavy weather and very lumpy seas, the crew hastily prepared to disembark. Two yawls were lowered and kept along-side. Into them went canvas bags containing coins and other valuables—"the results of their hazardous and wicked doings," whispered an 1895 account. In addition, each crewman secured about his person his share of the ship's spoils. Just before scrambling into the yawls the pirates attempted to scuttle their ship to destroy evidence. In their haste they failed. And not wanting the vessel to run up on a beach and incriminate them, they set her sails, lashed her helm, and headed her seaward into the teeth of the storm. Then all seventeen of them hurriedly piled into the yawls and rowed desperately for the beach.

Almost at once the boats were in trouble. So rough was the sea that they shipped water constantly and their occupants had to bail frantically for their lives. Now they were caught betwixt the devil and the deep blue sea. It was too late to return to their ship, now a derelict tacking into the gloom. Ahead, an angry surf thundered upon the beach. The freebooters loudly cursed their shoreside companion for beckoning them ashore in such a gale. Moments later both yawls struck a bar and were flipped over by waves, spilling all hands into the water. Shouts and screams rose above the wind's requiem as seventeen cutthroats splashed hysterically in a boiling sea. Only two made the haven of the beach. All the others, loaded down with plunder and unwilling to jettison it even to save their lives, went to the bottom like stones. Later their bodies were flung contemptuously onto the beach by breakers. All that night the two survivors, one Tom Knight and a Jack Sloane, aided by their burly shipmate who had awaited them on the beach, dragged the bodies in among the dunes and buried them in shallow graves in a deep hollow.

"Treasure Island"

Within a week, Captain Terry went on, Tom Knight and Jack Sloane appeared in various villages, spending money like . . . well, like proverbial drunken sailors. They were very generous, 'twas said, over-paying for even the most trivial purchases and always refusing change.

Rumors of dark and murderous deeds spread throughout the countryside, finally reaching such intensity that constables were dispatched to the beach to arrest anyone who might be in The Old House. All the lawmen bagged there were the stranger's loudly defiant wife and belligerent daughter. From behind a dune their paterfamilias and his two shipmates watched the constables drag the female wildcats off to the jail. Then the unholy trio hastened inside

33

The Old House. There each man unearthed his share of the loot where he had hidden it, then wended his way deep in among the dunes to bury it where he thought best.

In time Jack Sloane, Tom Knight, and their ugly companion, not identified by Captain Terry, dropped from sight, never to be seen in those parts or heard from again. So, too, vanished the big stranger's unpleasant mate and offspring.

Forty years later, the yarn has it, Tom Knight's gold was found, sealed in a black pot, where he buried it. No mention was made of its value. As for the booty interred among the dunes by the other two pirates, a romantic notion persisted for many years that it was still out there somewhere in the sand hills.

Probably fostered by rumors, word of buried treasure got around, and for a long time the beach where the yawls attempted to reach shore attracted several seekers of instant riches. Some Spanish coins were found—one lucky prober gathering thirty-eight of them, it was reported. Waves and tides permitting, searchers also dug into the sand bars believed to have capsized the yawls. Their labors, like those of most of the dune diggers, were in vain. A few coins and pieces of treasure bags were found, but nothing commensurate with time, effort and hope. If there's substance to the legend's talk of treasure, a fortune in coins and other valuables rests, probably forevermore, beneath the sand somewhere in Long Island's eastern reaches. Personally, I wouldn't want to bet on it. Besides, chances are that any treasure trove now lies underneath a shopping center, motel, or housing development.

An Unsolved Riddle

Out of sheer perversity, the pirates' ship survived the sou'easter as an unmanned derelict. In insane fashion she rode out the wild whimsy of wind and waves until finally, as though at the compulsion of contrariness, she did something her crew had feared. She came ashore, near Southampton. But she kept her secret, betraying no one. There were no papers to indicate her registry or owners, no cargo, nothing to indicate her origin or destination. Word was that a few Spanish dubloons were discovered in a locker, and that some pistols and cutlasses were scattered about. Nothing more was found when she was searched as she lay dead on the beach, said Captain Terry as he concluded his narrative.

However, an account several years after Captain Terry's story did mention this incident.

One night two bolder citizens of the region boarded her now-disintegrating corpse and proceeded to go through the wreck more meticulously than predecessors. Partway along in their search a sharp gust of wind whistled down through a companionway and blew out one man's lantern. He interpreted that as a threatening omen and promptly lost all interest in the project. His companion, a hardier soul, remained and was rewarded by a find of cached money. The story further has it that right afterward he and his family exhibited sudden and lasting prosperity.

Be that as it may, the mystery of the "Money Ship," as she came to be known, was never solved. It never was learned where she came from, who was aboard, or where she was eventually bound. Even her name remains a mystery. Nor was it ever learned what flag she flew. But if we subscribe to old Captain Terry's account we can assume that there were occasions when a Jolly Roger—a black banner with a white skull and crossbones—fluttered in her rigging.

5

Vessels from Hell

Of all calamities that befall mariners at sea, one of the most dreaded is fire. And two of the greatest coastal ship disasters in United States maritime history were caused by it.

Sixty-four years separated those two catastrophes, but both vessels were passenger-carrying steamers and nearly the same size. The steam packet *Lexington,* 220 feet long, was only five and a half years old when disaster overtook her on Long Island Sound between Connecticut and New York City one frigid night in January of 1840. The twenty-six-year-old excursion steamer General Slocum measured 280 feet in length and was destroyed by fire in New York City's East River near an expanse of water called—appropriately in this case—Hell Gate. Her date with destiny was the morning of June 15, 1904, and ironically she kept it within a stone's throw of shore.

Both disasters were incredible horrors. Each claimed a stunning price in human lives.

The "Lexington" Cremation

In addition to being quite young, the *Lexington* was well built and fast, one of the faster steam-powered vessels of her era. Her runs were between Connecticut and New York City, and speed was an attractive feature to people shuttling back and forth. There were no staterooms aboard the *Lexington*, but she did have a large, fancy main cabin which doubled as a comfortable lounge and dining saloon. Although she was little more than a ferry, she offered a pleasant voyage in miniature, a welcomed change from time-consuming, fatiguing land travel.

On the date of the fatal sequence of events, the packet was commanded by Captain George Child of Narragansett, Rhode Island. Under him was a crew of forty-one. A good number of passengers—men, women and children—came aboard from her wharf in Stonington, Connecticut, that January 13th afternoon. Despite the fact that it was midwinter, it should have been a routine crossing that began when lines were cast off at 3:30 P.M.

Outside the snug comfort of the steamer it was bitter cold, the thermometer slowly inching downward toward zero, then 10 below after sunset. Masses of ice floated in Long Island Sound, not large enough to menace the *Lexington*, but in sufficient numbers to make their presence known as they thumped and rubbed against her and reduced her headway. Below in her engine room the "black gang" fired her boilers to a maximum safe head of steam to offset the ice masses' slowing effect. Through the arctic night the *Lexington* pushed her way toward New York. In the cheery warmth of her grand saloon along about 7:00 P.M., the passengers were at peace with the world after a fine dinner, and looked forward to an evening of theatrical entertainment.

A Warning

The first threat of trouble came when the packet had chuffed only three or four miles off Eatons Neck on Long Island's North Shore. Wisps of smoke were noticed coming from around the lower sections of her funnel casing. Other plumes of smoke rose in the frigid air from bales of cotton piled on deck nearby. The ship's smokestack, overheated by extra firing of the boilers, had ignited adjacent woodwork. The smoldering wood, in turn, had touched off some cotton bales. The warning allowed for no debating. Fanned by a northerly wind, what might have been a controllable blaze if fought at once swiftly erupted in full-fledged conflagration, roaring and crackling beyond any efforts to extinguish it.

Captain Child headed his charge for the nearest Long Island shore at full speed, but the fire was outracing him, spreading with terrifying rapidity. In short time the *Lexington* became a torch. Fireworks of sparks shot into the night sky; flames leaped from the vessel, and smoke trailed astern. The nightmare worsened. Fire reached the steamer's tiller ropes, consuming them and leaving her completely unmanageable. The floating mass of flames yawed and swung crazily in Long Island Sound's ice-choked waters.

Attracted by an orange-red glow in the night, several people collected on shore. They were powerless to do anything at the moment, but it was thought that the burning vessel would reach proximity to shore within about fifteen minutes. Spectators made ready to give immediate assistance. In the interim, two boats rowed by eight men moved out into the sound to do what they could. Then fate stepped in to deal the *Lexington* another cruel blow. As the boats from shore fought their way toward the stricken packet, her rudder became jammed to one side, steering her in a great circle and exposing her to flame-feeding drafts from every quarter. The fire spread to more areas.

37

Even the wildest hopes faded. Captain Child endeavored to order passengers to the lifeboats, but panic reigned. The fire roared at its fiendish worst amidship, splitting passengers and crew into two frantic masses of humanity, fore and aft, and neither able to communicate with the other. Their hysterical milling, rushing and shoving deprived them of reason. Blood-chilling cries and screams fractured the quiet of the winter night.

A Crematorium

Only four small lifeboats were available. On Captain Child's order to take to them, the surge of humanity became even more frantic. In seconds the lifeboats were overloaded with terrorized, clawing people. One boat capsized immediately upon launching, dumping its occupants into freezing water. Other passengers, flames nibbling at their clothing or already consuming it, jumped overboard from the steamer. Silhouetted against a fiery backdrop, people ran every which way on the decks, torn between risking cremation if they remained aboard or chancing almost certain death in the icy sound. All this played to the accompaniment of an insane symphony of screams, cries and shouts. The scene could have been one from Dante's *Inferno*.

Nearest to this hideous drama were four men in one of the boats that had put out from shore. Somehow they had managed to work their way three miles out on Long Island Sound. Just as it seemed they might be able to assist in rescuing some survivors, there came another reversal by fate. As if conspiring with the conflagration in her own destruction, the *Lexington* perversely swung away from shore and started to drift faster. While the would-be rescuers watched in horrified frustration, the blazing funeral pyre floated farther and farther out, beyond all aid.

Within a few hours the *Lexington* was reduced to a charred, gutted, smoking hulk. Having consumed most of what they could, the leaping tongues of flame subsided, but fire continued its gnawing until the once-proud vessel's superstructure was completely gone. Hisses of steam were the only sounds breaking the bitter cold silence when the ravaged packet slipped beneath the waves.

The Toll and Loud Accusations

It could not be determined exactly how many people joined the *Lexington* in death. Contemporary estimates figured the toll at 118 to 141. Fire killed some; exposure claimed others. A belief was that drowning, along with submersion, destroyed the greatest number. Among the victims were Captain Child and several prominent

Bostonians and New Yorkers. Among the missing was an executive carrying $60,000 in cash for delivery to Manhattan brokers.

On the first list of *Lexington* passengers to be released to a stunned public was the name Henry Wadsworth Longfellow. By ironic coincidence, the poet had just sold his *Wreck of the Hesperus* to a New York newspaper (for the princely sum of $25, it was reported). It appeared in print the day after the disaster. Longfellow was supposed to have been aboard the packet that trip to keep a lecture engagement in New York; but his lucky star was shining, and a delay caused him to miss the fatal sailing.

Five men survived by clinging to bales of cotton that tumbled overboard. Eventually wind and currents carried them to safety far down Long Island's northern shore. One survivor was David Crowley, the steamer's second officer. Phenomenally, he weathered exposure to terrible cold for two days and nights before his bale-of-cotton raft came ashore near Riverhead, a town well east on Long Island. The other four survivors drifting their way on cotton bales were the *Lexington*'s pilot, Captain Stephen Manchester, a passenger from Norwich, Connecticut, one of the packet's firemen, and an unidentified man. Pilot, passenger and fireman were picked up by a sloop out of Bridgeport, Connecticut. The unnamed man was cast up on a remote beach near Hortons Point, Long Island, about fifty miles from the sinking. Half-dead from cold and exposure, he stumbled into a tavern in the village of Southold, east of Riverhead—an environment that may have helped speed his recovery.

The *Lexington* tragedy opened floodgates of violent criticism and grave charges. Magnifying the disaster was the fact that the public still viewed steam, as a means of ship propulsion, with suspicion. An explosion wasn't the cause; but for steam you must have fire, the dangers of which incited much outrage. Bitter, more logical attacks were leveled at carrying inflammable cotton on the decks of a passenger vessel, particularly near her funnel. Added were some very ugly charges. A Long Island newspaper termed the disaster "Willful, savage, horrid murder," and accused the ship's officers of acting only in their own selfish interests. The accusation went so far as to declare that the skipper was among the first officers to desert his post. He, of course, never had a chance to defend himself.

Newspapers were blatant vehicles of outraged cries. Preachers followed up from their pulpits. Poets were inspired to take up their pens. Artists were moved to special efforts too. One was young Nathaniel Currier, later to become senior partner in the famed lithography firm of Currier & Ives. His imaginative impressions of the *Lexington* disaster sold prints for parlors all over the country and started him on the road to renown.

Crass commercialism was inevitable, even then. Enterprising companies secured what bales of cotton had washed up on Long Island's coast from the packet and transformed them into "Lexington shirts" for sale as souvenirs.

An investigation probed the disaster. At its conclusion a coroner's jury placed the blame on steamboat inspectors for allowing the *Lexington* to sail with an inflammable deck cargo. The jury also censured the conduct of the lost vessel's officers. Captain Manchester, the surviving pilot, was absolved of any blame by the panel's fourteen members.

Years later Captain Manchester was to write a sequel to the *Lexington* story when a steamer he commanded met her end only a few miles from where the charred remains of the packet lay buried in 130 feet of water.

The "General Slocum" Horror

As huge and cosmopolitan as it is, New York City has always been composed of ethnic "villages"—that is, communities within the metropolis in which specific nationalities concentrated after coming from foreign countries. With the increasing tempo of immigration in the 1800s, a flood of humanity poured into the funnel of the Port of New York from Europe. In that rising tide came the Irish, Germans, Italians, Jews, Swedes, Poles and representatives of practically every country in Europe. Many kept going beyond New York, to join relatives or seek fortunes on their own in other parts of this lusty young nation. Many remained in New York City, settling in neighborhoods already established by countrymen. For many years these ethnic communities were very cohesive as immigrants preserved their native languages and customs as a buffer against the realities of life in a new and alien world.

In the early 1900s one such ethnic village-within-the-city sprawled across several blocks on Manhattan's Lower East Side. It was a very tightly knit community known as Little Germany.

A sizable percentage of Little Germany's residents contributed to parishioners of St. Mark's Evangelical Lutheran Church, whose pastor was Rev. George F. Haas. From the congregation and its families came the estimated 1,500 people (the exact number never was determined) who began to gather at about 8:00 A.M. on an East River pier at Third Street on the pleasant morning of June 15, 1904.

It was a red-letter occasion—the St. Mark's annual Sunday School outing. A band added to the festive atmosphere by tootling lively German airs. Rev. Haas, accompanied by his wife and daughter and assisted by another minister, shepherded the swelling flock in anticipation of a delightful sail upriver and a picnic at a place called Locust Grove, a popular spot for such outings, located just beyond Throgs Neck in the Bronx at the westernmost end of Long Island Sound. Most members of the happy, laughing flock were children, convoyed by their mothers and Sunday School teachers. It being a working day, only about forty men were in the big party.

Chartered for this great occasion was the 280-foot excursion steamer *General Slocum*. Built from wood and sporting three decks topped by twin funnels, she was a sidewheeler typical of that era's excursion steamer design. Commanding her was Captain William Van Schaick, the only skipper the twenty-six-year-old vessel had ever known. It's germane to our story to note that he was a captain with four decades of experience under his belt. As such he was respected, although his record seems to have been tarnished a bit by an alleged reputation for being accident-prone. Over the years the *General Slocum* had acquired a rather lengthy list of minor mishaps. However, there doesn't seem to have been anything serious enough to spoil her charter business.

"Cast off Lines!"

At 9:00 A.M., Captain Van Schaick at the wheel, the excursion steamer eased away from her Third Street dock and headed up the East River, to the accompaniment of voices raised in singing Martin Luther's hymn *Ein Feste Burg Ist Unser Gott* ("A Mighty Fortress Is Our God"). Already youngsters were scampering around the steamer's three decks, laughing and chasing one another. Their elders relaxed and watched the continuing panorama of the city, which is especially awesome when viewed from the water. In the ship's galley women volunteers busied themselves with readying clam chowder and other gastronomic delights for lunch.

Mercifully, there was no inkling that the fun would last little more than an hour.

At about 10:00 A.M. the *General Slocum* entered Hell Gate, suitably named because of its strong currents and periodic turbulence. Hell Gate is the East River's doorway to Long Island Sound.

41

Exactly how and where the fire started, and why it was not detected sooner, are unclear details. One account mentions it being seen from shore as the vessel passed Casino Beach, on the river's Long Island side. There a man thought he saw wisps of smoke. Whether it was already discovered aboard the steamer, we don't know; but detection couldn't have been long in coming, for the curlicues of smoke were quickly replaced by leaping tongues of flame. Only minutes before, band music had floated across the water. Now the sounds were shrieks of panic and fear. Other vessels spotted the fire and gave whistle blasts in warning.

Aboard the *General Slocum* the first alarm was sounded as she drew opposite 130th Street, Manhattan, perhaps not much more than a hundred yards from the nearest shore. Mothers and Sunday School teachers, assisted by the relatively few men in the large gathering, hastily rounded up their young charges to find places of safety on the steamer. In short order there were none. Her cruising speed increased by a strong tide, the *General Slocum* moved into a brisk breeze that fanned the flames and sent them eating their way toward the stern. Hundreds of children had been marshaled aft. Only moments earlier the stern appeared to be the safest place. Now a head-on wind drove roaring flames toward them. This was a rapidly spreading fire. The wooden steamer quickly became a mass of flame.

Later it was learned how pathetically the *General Slocum* was equipped to cope with such an emergency. Her fire hoses, deteriorated through years of disuse, ruptured under water pressure. The lifeboats were so secured in their davits that they couldn't be launched. Adding to the confusion and a lethal delay, many members of the twenty-three-man crew were untrained. It subsequently was charged that they endeavored to save only themselves.

With fire hurrying toward them from every direction, the passengers became a terrorized, panic-stricken mass. Flames already had killed many; others lay unconscious from smoke inhalation. Numbers of them leaped overboard from the three decks, their clothing ablaze. Others ran about aimlessly or stampeded, not knowing what to do. It will never be known how many small children were trampled or crushed to death, or how many passengers were fatally injured by falls and collapsing superstructure. Those who found life preservers had them come apart in their hands.

Fatal Judgments

What of Captain Van Schaick while all this was going on? His activities can be pieced together only from testimony given later.

It seems that the captain turned the wheel over to a pilot named Van Wart when the fire was discovered, giving him orders to beach the ship on North Brother Island. On the face of it, this was a sensible command. As it worked out, however, North Brother Island was a poor choice. To begin with, it drew the *General Slocum* into a strong breeze out of a northerly quarter. Further, it was a long way around to safety. Swinging her rudder hard aport would have beached her within a few minutes, at the same time lessening the spread of wind-fanned flames, and would have provided a good opportunity to get many of the passengers off.

Instead, the *General Slocum* headed at all available speed into a stiff breeze and crashed into a rocky section of North Brother Island at a location where its rocky shore shelves steeply into deep water. The reason for this calamitous error in judgment was never explained.

Vessels moving in on the scene and spectators gathering on shore witnessed a hideous drama. The uppermost deck already had collapsed, hurling passengers into the water or dropping them to a hungry blaze below. Horrified watchers saw or heard people leaping from the doomed vessel or being trapped. When the *General Slocum* struck the North Brother Island rocks, another deck and more superstructure collapsed in flames.

The floating inferno finally was beached at Hunts Point. What was left of the hapless steamer burned to the waterline, completing her destruction, and cremating an untold number of persons.

The Terrible Tally

Since it was only a one-day excursion and so many people were expected, no passenger list for the outing had been drawn up. Therefore, it became impossible to arrive at an accurate total of victims. Knowledgeable estimates agreed, however, that more than 1,000 passengers—chiefly children, of course—were burned to death, drowned, or killed by other causes. Among the victims were the wife and daughter of Reverend Haas. Most of those who had leaped from the vessel to escape a fiery death met an end by drowning. In the early 1900s, unlike today, very few women and youngsters learned to swim.

There were approximately 407 survivors, many of whom were disfigured by burns or crippled for life.

Medical aid had been given at an emergency field hospital hastily set up at riverside at the foot of the city's East 138th Street, and it undoubtedly saved a number of lives. Bodies were brought to this location too, and later that day were transported by boat to an East 26th Street dock where they could be taken to Bellevue Hospital for identification by relatives. Even hardened newspapermen covering the tragedy became nauseated when they saw the victims. Some broke down and wept unashamedly.

Bitter Condemnation; Praises Too

President Theodore Roosevelt personally ordered an investigation of the *General Slocum* disaster. Some of the probe's findings defy belief. Extremely grave charges were issued against the captain, crew and others involved in either operating or equipping the excursion steamer.

One of the charges was leveled at the untrained seamen and their incompetence. To compound that crime, they also stood accused of cowardice. It was reported that most of them considered only their own safety and swam ashore when the *General Slocum* struck the rocks at North Brother Island. Only one crewman was recorded as a drowning casualty. At that time, it also was alleged, Captain Van Schaick, pilot Van Wart, and another pilot named Edward Weaver leaped to safety on the deck of a tug that had drawn alongside.

Findings indicated that the fatal fire probably started in a cabin where oil and other flammable liquids were stored; but how it actually began couldn't be determined. Some probers theorized that an exploding cook stove might have been the villain, but this was never proved or disproved. Nor was it ever explained why the fire was not detected sooner, while it still might have been possible to contain it.

On the other hand, maybe it wouldn't have been possible to contain a blaze even in the initial stage. The cruel facts were that the steamer's lifesaving gear was in deplorable condition; the untrained crew's incompetence undoubtedly contributed to the tragedy; and the lifeboats couldn't be launched. Further, it was claimed in court that the life preservers' manufacturer had secreted pieces of iron inside the preservers to bring them up to the legally prescribed weight!

The *General Slocum*'s fire-fighting and lifesaving equipment had been inspected—and approved!—only a short time prior to the disaster. For this negligence a supervising inspector and two underlings were fired. No law covered the chicanery involving the life preservers, but the Department of Justice indicted four employees of the cork company manufacturing them.

Now we come to the future of Captain William Van Schaick, then sixty-one years old. Except for minor burns, he had escaped unscathed—physically, that is. Police arrested him, pilots Van Wart and Weaver, and as many crewmen as could be located. Along with the managing directors of the firm owning the *General Slocum*,

6

A Femme Fatale Named "Mary Celeste"

Until the Archangel sounds his (or her) trumpet and the sea gives up its dead, the fate of the captain and crew of the brigantine *Mary Celeste*, together with that of the master's wife and daughter, will remain among the most intriguing mysteries of all time, on sea *or land*. All kinds of theories and purported "true accounts" have been offered to solve the riddle; but—and this seems highly improbable now—unless uncontestable evidence finally comes to light, the *Mary Celeste*'s grim secret will remain locked in King Neptune's confidential files.

Background of an Enigma

The *Mary Celeste* began life under another name. As the *Amazon* she took form in a shipbuilding yard in Parrsboro, Spencer Island, Nova Scotia. Launching, with appropriate festivities, came on May 18, 1861, the year civil war erupted in the United States. Launching also was attended by some difficulties, caused by the yard's balky ways. Years later, in retrospect, this was interpreted as a bad omen. No one worried about it at the time, but they may have begun to wonder not too long afterward. The *Amazon*'s first captain took sick and died before completing an early voyage to Portland, Maine, with a cargo of fertilizer.

As a brigantine—also called a half-brig—*Amazon* held two masts for her canvas. Her foremast carried squarish sails; while the mainmast held a fore-'n'-aft type of sail. The usual brigantine plan called for a gaff-rigged tops'l over a mains'l on the mainmast, but the *Amazon* was outfitted with just an oversized mainsail. In all other respects she was rigged conventionally.

Amazon would be called a dwarf in comparison with modern ships. She was even on the small side compared with some of the tall ships of her own era. In overall length she measured just under 100 feet, stem to stern. On the outside she had a width or beam of twenty-five and a half feet. Her registered tonnage was 198.42.

She started her career quite well. Her troublesome launching and the demise of her first master faded into the background as she made a number of modestly profitable voyages under the British flag. At least they were profitable enough to satisfy her Nova Scotian shareholders. Quite an ordinary looking vessel, she sailed in and out of U.S. Atlantic Coast ports and harbors in England without impressing anyone.

After half a dozen years of colorless but faithful service, during which time she was commanded by a series of captains, the *Amazon* experienced her first serious brush with fate. In September of 1867 she put in at Halifax, Nova Scotia, where yet another master took over. Acting on orders from her owners, the new skipper sailed his charge into the tricky waters of Big Glace Bay near Cape Breton Island. These waters were known to be dangerous in heavy weather, and September was bad-weather time in the area. Its nor'easterly gales had a disastrous way with small ships; and when the *Amazon* was battered by one, she was driven ashore in a place that made salvage difficult and costly.

Records state that the *Amazon*'s owners were unable to collect any insurance from the mishap because her policy was not binding in the area where she was stranded. They did nothing to salvage her, and her first British registry was terminated.

In all probability, the brigantine was not too severely damaged by her encounter with calamity, for soon she was re-registered in Sydney. Although no copy of this registry certificate is extant, records indic‧ ‧e that she was put back on the ships lists on Novembe‧ ‧1867, with permission and approval of Nova Scotia's lieuten‧ ‧vernor.

Under ‧ New Name and a New Flag

‧ ‧ time after she was re-registered, *Amazon* cruised into a U ‧ ‧rt on the Atlantic seaboard. There she dropped her second B‧ ‧n registry and was listed on the U.S. merchant marine service roster. New ownership would seem to be the reason for these changes, but details are hazy at best, so the new phase in her career becomes one of the missing pieces in the puzzle of this strange vessel. Like many of her other particulars, it's a detail that remains a secret of time.

What may have been the most notable alteration in her career occurred at the time her registry shifted to the United States. It was then that her name was changed from *Amazon* to *Mary Celeste*. (If she was named for a lady, her identity has long since been forgotten. Probably she was named for an owner's relative.) Old-time mariners were filled with superstitions, in which regard they felt very strongly about names of vessels. Many believed that it was very unlucky to change a name. They might very well have been right in

48

this instance. At any rate, the brigantine became so well known as the *Mary Celeste* that few people realize she had another name.

Even after she became a United States vessel, the rechristened *Mary Celeste* was not yet done with changes. Re-registered no less than five times between December 1868 and October 1872, she acquired a procession of owners, named Winchester, Goodwin, Samson, and Briggs. That last gentleman, Benjamin S. Briggs, was to be remembered for disappearing from her without a trace.

In the autumn of 1872 the brigantine underwent major physical alterations. Her specifications were increased to a length of 103 feet overall, with a twenty-five-foot-seven-inch beam, a depth of sixteen feet, two inches, and a tonnage of 282.28. Later, experts were puzzled as to how the extra length was added.

There's a hiatus in records of the *Mary Celeste* for the period between her alterations in the fall of 1872 and her departure from New York that November 7th. What is more important to the case, not even fragments of information gave clues to her activities between that departure date and the fifth of December, 1872, when she became the star of one of the most fascinating sea mysteries of all time.

Her Master

Benjamin S. Briggs, short-bearded, level-eyed, with a reputation for being as good-natured as he was ruggedly handsome, was a captain who hailed from a family of seafarers. The son of Captain Nathan Briggs, he was born in the village of Marion on the shore of Buzzards Bay, Massachusetts. Benjamin was one of six brothers, five of whom followed the sea. He and his brother Oliver both acquired master's papers at an early age.

Prior to the *Mary Celeste*, Benjamin Briggs commanded the brig *Sea Foam* and the schooner *Forest King*. There's every reason to believe that he was a thoroughly competent, reliable skipper. When he relinquished command of the *Forest King* to his brother Oliver, he had no trouble at all in securing another berth as master of the bark *Arthur*. Becoming part owner, as well as captain, of the *Mary Celeste* was a step up the ladder of maritime success.

Everything in his background pointed to Benjamin S. Briggs being a most worthy master of the brigantine. If there was anything suspicious about the *Mary Celeste*, as some observers hinted later, it lay with the vessel herself, not with Captain Briggs. That he had the utmost confidence in his own ability and his vessel was proved when he took his wife Sarah and young daughter Sophia aboard on the first voyage under his command.

The Voyage to Immortality

The *Mary Celeste* finished lading her cargo, 1,700 barrels of alcohol, on November 2nd. Brigantine and cargo were insured, and clearance was obtained from Port of New York authorities. Captain Briggs wrote a cheerful letter home, expressing regret that he didn't get to see brother Oliver before sailing.

Next day, he signed on a crew for the voyage. A man named Albert Richardson was hired as first mate; Andrew Gilling signed as second officer. Filling out the crew were cook-steward Edward Head and four men of German extraction: Boy Lorenzen, Volkert Lorenzen, Arian Martens, and Gottlieb Gottschalk (or Gotschall). Apparently the skipper was satisfied with this selection, for Sarah Briggs wrote in a letter home that he "thinks we have got a pretty peaceful set this time all around, if they continue as they have begun." It's to be wondered if Sarah's words "this time" meant that Captain Briggs had trouble with a previous crew. If so, perhaps it's a clue to a change in his personality at sea, which could have had bearing on subsequent events. Or maybe there's a key to the mystery in Sarah's phrase "if they continue as they have begun."

With his vessel in trim, cargo stowed, crew at their chores, and his wife and daughter comfortably settled aboard, Captain Briggs was ready for sea. The *Mary Celeste* slipped her New York mooring on Tuesday, November 5, 1872.

In short order the brigantine encountered a temporary setback. Because of suddenly adverse weather and sea conditions, Captain Briggs decided to be cautious. He dropped anchor off Staten Island to wait out the weather. On November 7th, a pilot named Burnett was taken aboard and *Mary Celeste* passed through The Narrows under his direction. Clear of that busy passage, the pilot was dropped off the ship. With him went a final missive from Sarah to her family in Massachusetts. The brigantine set a course for the Strait of Gilbraltar, the Mediterranean, and Genoa, Italy.

A Discovery

Nothing is known about the *Mary Celeste* and the people aboard her during the next four weeks, other than that she was progressing toward her destination. We can only assume that everything was in order—for a while, at least.

On December 5th, one month to the day the *Mary Celeste* left her New York mooring, the British brigantine *Dei Gratia*, Captain David F. Morehouse commanding, was nineteen days out of New York, Gibraltar-bound with cargo, and slicing through the Atlantic between the Azores and Cabo da Roca, Portugal. With Captain Morehouse on the quarterdeck was first officer Oliver Deveau.

Having sighted another brigantine, and perhaps even recognizing the cut and set of her sails as belonging to the *Mary Celeste* under command of his friend Ben Briggs, Captain Morehouse altered course to bring his vessel within hailing distance. When his shouts brought no answering hail, and sweeps of his glass revealed no signs of life aboard, he ordered a boat away to investigate.

In the boat which pulled toward the *Mary Celeste* were the *Dei Gratia*'s first mate and two seamen named Anderson and Lund. Although a fairly high sea was running, the small craft was able to come alongside the apparently deserted ship, and the three men boarded her.

They searched her fore and aft, topside and below, even her holds. After an hour or so, the boarding party returned to the *Dei Gratia*, where first mate Deveau reported that there wasn't a soul aboard the other ship. She was deserted and derelict. Following custom, Captain Morehouse sent his first officer and skeleton crew to the *Mary Celeste* with orders to sail her the remaining 600 miles to Gilbraltar.

Puzzles Everywhere

If the *Mary Celeste* had been in a battered or otherwise unseaworthy condition when found by the *Dei Gratia*, there might have been no mystery about the complete disappearance of all hands. Rightly or wrongly, it then could have been assumed that the brigantine had encountered a violent storm or particularly vicious seas, bad enough for Captain Briggs to fear that his ship would founder and to consider emergency abandonment. A natural sequence of logic, then, would have been that all hands met their end in the *Mary Celeste*'s yawl, which was missing. It would have had to have been a severe emergency for a seasoned master such as Benjamin Briggs to abandon a larger vessel and risk the lives of all concerned in a cockleshell. Or possibly he had made a fatal error in judgment.

Such things might have been conjectured if the derelict had exhibited serious storm damage. The catch was, she hadn't. Quite the contrary. When boarded by the *Dei Gratia* party she was in good trim, entirely seaworthy and navigable, as was proved by the skeleton crew sailing her into Gibraltar without difficulty. Her good condition was sworn to by first mate Deveau.

Had there been any signs of violence, any at all, they might have provided a tangible clue to what had transpired on the *Mary Celeste*. Perhaps a mutiny could be speculated, prompted by some real or fancied wrong, in which the rebels disposed of their captain and his family, then aborted their plans out of fear and fled, preferring to take their chances with the Atlantic. Any signs of violence aboard might also have pointed to barratry—theft of the cargo of alcohol by the crew, in this case—or an encounter with a nomadic pirate. Pirates had become rare by that time but were by no means completely out of existence.

Here again investigators ran into large snags. There were no signs of violence, bloody or otherwise. Further, her cargo appeared to be intact and nothing was missing from the known personal belongings of Captain Briggs, his family, and the crew. If there had been an attempted mutiny, barratry or piracy, the rogues would have looted at least as much as they could carry, and most probably would have scuttled the vessel to forever conceal evidence.

However, the *Mary Celeste*'s yawl was gone, and the *Dei Gratia* boarding party could find no sextant, ship's chronometer, navigation books, registry certificate or ship's papers. Their absence deepened the mystery.

Another contemporary theory echoed a belief popular among seafarers of that era: That the *Mary Celeste* had been attacked by a giant octopus or a huge squid which had attempted to drag down the vessel and, failing in that, seized the people aboard her. The greatest flaw in this speculation is that no *known* species of octopus grows that large, and even the giant squid, *Architeuthis*, is an unlikely suspect since it resides at great depths (1,500 feet or more). Besides, any monster of such enormous size would have inflicted some damage on the vessel. Furthermore, it's highly unlikely that everyone would be on deck simultaneously at such a time; and, even if they were, it's even less likely that they would have obligingly allowed themselves to be plucked like fruit without struggle.

Today, someone would surely suggest that they were captured by a UFO. The existence of UFOs was suspected long before the *Mary Celeste*'s day, but such speculations were not voiced in 1872—not out loud, anyway.

Last, but far from least, the derelict brigantine bore no signs of fire, explosion, collision, or hull damage from unknown causes. However, her volatile, highly inflammable cargo of alcohol has figured in an interesting speculation.

Perhaps something happened to that cargo that made Captain Briggs fearful of fire or an explosion. For example, one or more barrels of alcohol could have ruptured, filling the hold or the entire vessel below deck with combustible, explosive fumes. Fearful lest these fumes be touched off at any time, the skipper could have issued orders for immediate abandonment of the brigantine.

Carrying this speculation further, all hands left hastily in the yawl, withdrawing to a safe distance to await developments. When none were forthcoming, they decided to return to their ship. Just then a breeze suddenly sprang up, causing the *Mary Celeste* to move away. Gaining speed, the ship soon widened the gap between herself and her crew. The yawl was left behind on a wide ocean, without food and water, since her occupants had departed in great haste. While ten souls watched the delivery of their death warrant, their ship disappeared over the horizon and into immortality as one of the sea's great riddles.

Second-Guessing

"Solutions" in profusion have been offered in vain attempts to solve the *Mary Celeste* riddle. Additionally, at intervals there have appeared so-called true stories of what actually happened. But in all these instances certain flaws and discrepancies, major and minor, have popped up to discredit them. The pertinent details, gleaned from testimony given at a British admiralty court of inquiry at Gilbraltar, are as follows:

At the time of discovery, when the *Dei Gratia*, sailing in comparatively rough water on a course southeast by south, came across the *Mary Celeste*, the derelict was running northwest by north at about two knots on a heading about 180 degrees out of the one she was supposed to be on. Although she flew no distress signals, she appeared to Captain Morehouse to be in some kind of difficulty. Her fores'l and upper fore-tops'l were gone, and the main stays'l was piled on her forward deck. The lower fore-tops'l was set but torn; a jib and fore-topmast stays'l were set in good order. Other canvas was furled. No aftersails were set. The vessel was on a starboard tack; and although no one was at her wheel, she didn't yaw wildly.

The *Dei Gratia* headed a point or two into the wind and waited for the *Mary Celeste* to close the gap. When the two vessels were close enough to each other, Captain Morehouse hailed her.

As already mentioned, her hull was in fairly good shape. Some water in her hold lapped at the barrels of alcohol, but it wasn't a dangerous amount and it was easily sucked out after her pumps went to work. All masts and spars were up, although parts of the standing rigging and running rigging were out of commission, with some sheets and braces over the side. Anchors and their chains were all in place. Two or three hatch covers were off and lying on deck. Her water casks were intact in their chocks. The binnacle lay on deck near the wheel, its compass broken, ostensibly when it fell. The steering wheel was operable and had not been lashed. Six deck cabin windows had been sealed with canvas and boards, seemingly for heavy weather.

A companionway door and galley door were ajar. Although there was some water in the galley, the result of waves breaking on deck, everything was in order there. All cooking gear had been put in place. There were no signs of food preparation (this is in opposition to certain accounts, which also mention a teakettle "singing" on the stove). All was in order in the forecastle, the crew's quarters. Its four berths were neat, and the sailors' sea chests contained extra clothing, foul weather gear and personal possessions. None of the items a seaman would hastily grab if he were abandoning ship seemed to be missing. And the brigantine carried an estimated six months' supply of food.

The main cabin, whose door was ajar, was wet but without standing water. A clock had been ruined by water, probably coming in through a hole in the skylight. The mate's deck log lay on a table. It had entries up until 8:00 A.M., November 25th, by which time she had passed to the north of Santa Maria Island in the Azores. A rack for keeping dishes in place also was on the table. There was no food about, or any sign that a meal had been in progress.

Captain Briggs' cabin, abaft the main cabin, contained a melodeon, or small reed organ, two boxes of clothing, a few sewing articles, a scattering of toys, several rolled-up navigation charts, and some books. The last entry in his log was dated eleven days earlier, November 24th. The cabin's bunk was unmade and looked as it would if someone had just arisen from it. Under this berth lay a fancy Italian sword bearing what appeared to be blood smears— later disproved as such, apparently.

On a desk in first mate Richardson's quarters were the ship's log with entries up until November 24th, two charts with the *Mary Celeste*'s course plotted to the 24th, a cabin compass, a notebook, and a kit of tools.

Note that date, November 24th. Whatever happened, it occurred between that date and December 5th, most likely either later on in the day or on the night of the 24th or 25th.

The *Dei Gratia*'s first mate, Oliver Deveau, testified at the official inquiry that everyone aboard the brigantine seemed to have abandoned her in haste, yet there were no indications of a struggle or violence. He repeatedly stressed that there was nothing really wrong with the vessel. He mentioned finding two quadrants, adding that no sextant—vital to navigation - or chronometer, navigation books, registry certificate or vessel's papers could be located.

We'll never know its true significance, and under the circumstances it's little better than no clue at all, but the missing yawl has to be a key to the disappearance of the *Mary Celeste*'s people. There were no davits on the brigantine's port and starboard sides, and her stern davits had a spar lashed across them. Indications were that the missing yawl had been secured atop the main hatch. *Someone* must have left in that craft.

Disposition of the yawl was only one of several missing pieces in the puzzle. Another was how had the *Mary Celeste* managed to hold a course with no one at her helm and the wheel not lashed? Still another, why were the hatch covers off? Had Captain Briggs removed them to allow combustible fumes of vaporized alcohol to escape, or was it possible that the fumes had built sufficient pressure to blow them off? We keep coming back to that cargo as another potential key to the mystery.

Closing Chapters

The admiralty inquiry in Gibraltar shut its book on the *Mary Celeste* case without reaching any definite or satisfying conclusions concerning the fate of Captain Benjamin S. Briggs, Sarah Briggs, Sophia Briggs, and the seven others. A salvage fee—a fair amount for those days, but a fraction of the worth of the vessel and .'r cargo—was paid to the *Dei Gratia*, and that was that for the time being.

For a while the mystery brigantine was a curiosity at quayside in Gibraltar. Her case already had the peculiarly potent attraction it has held ever since.

Under command of a replacement skipper sent from the United States, the *Mary Celeste* delivered her cargo to Genoa as originally scheduled, then returned to her home country. For several years thereafter she sailed to and from various ports, earning money but without notable financial success. She also saw several changes of owners.

In Boston in December 1884, an aging, now rather unkempt-looking *Mary Celeste* took on a cargo for transport to the Greater Antilles in the West Indies. Her destination was Port-au-Prince, Haiti. Under command of a Captain Gilman C. Parker, she put to sea with a heterogeneous collection of freight rather heavily insured by five different companies.

The Jonah that had accompanied her since launching difficulties twenty-three years earlier was still aboard. *Mary Celeste* never reached Port-au-Prince. She met death on Rochelois Bank, a coral reef in Haiti's Gulf of Gonave.

When an insurer became suspicious, Captain Parker and others were brought up on charges of collusion to commit barratry, which means fraud and/or criminal negligence by a ship's officer and crew against owners or insurers. Particulars of the charge at their trial in federal court in Boston were that the defendants took on a cargo of appreciably less value than declared for the vessel's manifest, then purposely wrecked the ship. A deadlocked jury ultimately discharged the defendants, but the "murdered" *Mary Celeste* had a revenge of sorts. Among those involved, the trial's aftermath included a suicide, a defendant who went insane, and a bankruptcy.

The *Mary Celeste*'s brutal demise on a Haitian coral reef was a sad end, yet strangely in keeping with her unhappy, tragic life. It all happened long, long ago, but the case of the strange derelict seems destined to sail on forever, to be thought about and talked about whenever maritime mysteries are discussed. A Gibraltar court of inquiry may have closed the official dossier on the ill-starred brigantine, but until the Archangel sounds the trumpet and the sea gives up its dead, there will be no final words to the mysterious story of a *femme fatale* named "Mary Celeste."

7

Tragic Saga of the "Sea Dragon"

The great adventure of the *Sea Dragon* is not a maritime mystery. How she vanished with all on board in the Pacific Ocean somewhere west of Midway Island has been established without question. Nevertheless, hers is a very unusual story, and there's something strangely disturbing about her first and last long voyage.

Perhaps the case of the *Sea Dragon* is disturbing because long after she vanished there were still optimists who entertained hope that somehow her people had managed to reach a remote atoll and survive*. Or maybe it's because wreckage, thought possibly to be the *Sea Dragon*'s remains, but never proved, was reported at intervals up until six years afterward. There was a sighting at sea by a U.S. passenger liner en route home from Yokohama. Unfortunately, because of rough seas, she was unable to bring the wreckage aboard. Five years later, a length of wooden keel with some ribs attached, such as might have been part of the *Sea Dragon*'s anatomy, washed ashore on a California beach. In both instances it was calculated that currents could have transported the wreckage to the places of discovery in the time involved, but nothing conclusive came from either instance.

With or without mystery, the *Sea Dragon*'s story stands as a fascinating one.

Her Owner

Even with stiff competition from a galaxy of the brightest stars in Hollywood's film firmament, it would have been difficult in the late 1920s and early '30s to find anyone who enthralled the American public more than Tennessee-born Richard Halliburton. Here was a fellow gifted with enormous magnetism for both sexes. To women he appeared good-looking, virile, charming, and, according to their preferences, urbane or appealingly boyish. To thousands of males, particularly Walter Mitty types chained to lives of humdrum

---

*A few of these optimists may yet be around. In an eerie sidelight, for more than twenty years afterwards, letters were being addressed to her owner as though he was still alive.

routine, he was the fearless vagabond-adventurer they would have given their teeth—and maybe their wives as well—to be.

Wild, daring exploits around the globe were Richard Halliburton's stock in trade. He was possessed by a strange, all-powerful drive to accomplish one difficult or fantastic feat after another. He climbed the forbidding Matterhorn; swam the Panama Canal—the hard way—its full length; lived among some of the world's most dangerous convicts in the infamous French penal colony on Ile du Diable (the dreaded Devil's Island) off the northeastern coast of South America; dove seventy feet into a Mayan "well of death" deep in the jungles of Mexico; retraced Hannibal's route across the Alps, like his famous predecessor, on an elephant; swam the turbulent Hellespont (Dardanelles) between Europe and Asia Minor, and marched with the French Foreign Legion, among other feats.

If there was no challenge at hand at the moment, Richard Halliburton went out and found one. He was forever finding ways to risk his life. In retrospect, it was an accomplishment *per se* that he lived as long as he did—and he didn't live terribly long.

Admittedly, there was a method to his madness. Halliburton shrewdly realized that there were thousands of people everywhere who hungered for the excitement of adventure, but who would never experience it personally; and he had a superb talent for giving them what they craved "by proxy"—in print. He wrote absorbing books about all his exploits, and from the very first volume, *The Royal Road to Romance,* and the series which followed, they were bestsellers. Halliburton's books eventually were translated into some fifteen foreign languages and copies were bought in the hundreds of thousands by an insatiable and adoring public. He also took to lecture circuits, and at the peak of his popularity packed halls to the rafters everywhere. In between he turned out a profusion of magazine articles. Much of the proceeds from all these endeavors went to finance still more adventures, providing material for more books, lectures, and magazine articles. Richard Halliburton became one of the world's best known—and most envied—men. He was riding high on waves of idolization, fame, and money.

But there was trouble in this paradise. Halliburton was aware of it. No one knew better than he that even the most ardent admirers can become jaded in time. Halliburton found himself in the difficult position of having to come up with progressively more spectacular and dangerous encores.

His Stage Is Set

A feared decline in public interest became evident toward the mid-1930s. Sales of his books started to skid downward, and attendance at his lectures fell off noticeably. Halliburton desperately needed a kind of super-adventure.

And so he was ripe for one of his most perilous ideas: He would go to the Orient and sail a Chinese junk across the Pacific Ocean. The idea in itself appealed to him enormously, and it did spawn interesting commercial possibilities. As he viewed the plan, he would time his arrival in California for San Francisco's Golden Gate Exposition in 1939. A transoceanic voyage in a junk should restore his fame to full flower while providing exciting material for new writings and more lectures. It also could be profitable from another angle. Arriving in San Francisco in a blaze of glory, his voyage having been publicized along the way, he would put the junk on exhibition at the Golden Gate fair and charge a fee to board her. In a letter to his parents dated August 31, 1938 from San Francisco he wrote: "I'm going to have the most exciting concession at the fair."

Halliburton journeyed to Hong Kong. In the Kowloon shipyard of one Fat Kau he ordered construction of a Ningpo-design junk, seventy-five feet long and twenty feet in the beam—"extra sturdy and adorned with every possible flourish," he wrote in his notes. That was in the autumn of 1938. On November 2nd he commented, "Our junk will be finished in forty-five days. Fifteen days more to rig and train our crew. Then we plan to set out about January first." It turned out that he was overly optimistic in his timetable. He hadn't reckoned with the ancient, frustratingly slow handcrafting methods of Chinese junk builders.

Courage or Madness?

Halliburton was sold on junks and considered them very seaworthy. Such enthusiasm was not widespread outside China. Certain types rode too high in the water, with a drift that was difficult to control. Other junks rolled alarmingly. Steering was primitive, just a wooden tiller, with a distinct possibility of heavy seas snapping off the massive rudder if construction was not all it should be. Halliburton countered these and other criticisms with a reminder that the Chinese had been sailing around in the ungainly arks for centuries. He saw no frightening difference between mariners who had developed a kind of inborn junk-navigating ability and a crew, chiefly non-Chinese, that would have to be trained, and in only a couple of weeks at that.

To most observers, sailing one of those awkward-looking vessels across *any* ocean, not to mention one known as a breeding place and playground of fierce typhoons, sounded like a sure route to suicide.

Some were of the opinion that it was Halliburton's craziest stunt yet and said so, adding that he'd never make it.

Storms and peculiarities of junk behavior were not the only hazards. At that time Chinese pirates, driven south by Japanese naval operations, swarmed in the waters around Hong Kong—they even attacked the Hong Kong-Kowloon ferry one night—and along Halliburton's proposed route in the Formosa Straits. Moreover, China and Japan were at war, and ever since a gunboat hidden among some innocuous-looking junks had darted out to torpedo and badly damage a Nipponese aircraft carrier, the Rising Sun navy tended to sink Chinese junks on sight. Halliburton figured that documents from the Japanese Foreign Office and an American flag prominently displayed on the junk would provide protection enough. This confidence had very fragile backing. Japan was girding itself for a far-flung Pacific war; and Halliburton had not always spoken kindly of Japan. So far as Chinese pirates were concerned, his voyage would be strictly his own. Against them he planned to carry an armament of rifles and shotguns.

Numerous factors gave the master adventurer pause for thought, but none deterred him. Nor did the efforts of his parents and friends to dissuade him. Nor did the loss of famed aviatrix Amelia Earhart in the Pacific during an around-the-world flight only months earlier. Once Richard Halliburton set his sights on a goal, there was no turning back. However much he may have worried privately, adversity seemed to add fuel to his enthusiasm. He shrugged off as superstitious claptrap a solemn pronouncement by senior Chinese observers that naming his vessel *Sea Dragon* would handicap her with an evil omen. A dragon is a *land* animal, they pointed out, and said that to name a vessel for one was to deliberately invite misfortune. To offet this, Halliburton could argue that his junk would have her stern decorated with a painting of a large phoenix, a mythical bird considered a symbol of good luck. He had a few superstitions of his own.

"One More—
One Last—Good-bye . . ."

The workmen at Mr. Fat Kau's establishment labored with the velocity of snails much of the time, and there were maddening delays in construction. But in time her owner's mounting impatience was salved by the *Sea Dragon*'s shakedown trials. She moved right along when the wind filled her gaudily colored sails, but her rudder was cumbersome and required two men to manage. Her 100-horsepower diesel auxiliary engine, installed as an afterthought, performed satisfactorily. Leaks were stoppered, and a fin-keel was added in drydock to lessen her disconcerting tendency to roll.

It finally appeared that the Oriental lady was ready for her grand debut on the Pacific. She would be under the command of Captain John W. Welch, a ticket-holding master from Australia. Also engaged were an engineer, radioman, and a polyglot crew of Chinese, and professional, part-time Occidental sailors. Halliburton would pitch in as a crewman too. Originally, a few of his friends from the States were to be aboard; but they lost enthusiasm during preparations. Two cancellations came practically at the last minute. One fellow was still confined to a local hospital after surgery, where he told Halliburton he had changed his mind and wouldn't have gone anyway. The other friend resigned out of trepidation. Theirs was a fortunate decision. Lucky too were the people back home who begged to be taken along—hundreds volunteered—but had to be refused.

The *Sea Dragon* was laden with provisions for three months, 2,000 gallons of fresh water, and enough fuel for twenty days' propulsion by her diesel engine. Based on a trial run in February of 1939, Halliburton figured that they would cover "170 miles a day . . . 1190 miles a week . . . 8000 miles in less than seven weeks. We'd be home by the end of March!" But he was unknowingly prophetic when he began what would become his final communiqué to his parents home in the States: "One last—good-bye letter. We sail . . . in a few hours . . ."

Outward Bound

In fair weather on March 5, 1939, the *Sea Dragon* lay quietly poised for departure. She looked splendid, glistening in spanking new paint, standing out with her elaborate colorful decorations, and protected by the traditional two large "eyes" carefully positioned on a junk's bow in the belief that they would enable her to see where she was going at all times. Spectators crowded the docks and bade her bon voyage in assorted dialects as she pulled away.

The *Sea Dragon* cleared busy Hong Kong harbor and stood out into the South China Sea. No problems with Chinese pirates or Japanese war vessels. She was on her great adventure at last. Halliburton's exhilaration must have soared. Right up until sailing time he maintained a personal log of the expedition's progress. He undoubtedly continued scribbling during the voyage, seasickness permitting. We'll never see those notes. We can follow the cruise to oblivion only through exchanges of radio messages.

This one sped through the ether on March 13th: JUNK SEA DRAGON VIA SAN FRANCISCO . . . 1200 MILES AT SEA ALLS WELL.

By now the junk had dropped Okinawa astern and was making her way south of Japan.

From her came daily progress reports, intercepted regularly. On March 19th, she radioed: JUNK SEA DRAGON VIA SAN FRANCISCO . . . HALFWAY MIDWAY ARRIVING THERE APRIL 5TH SKIPPING HONOLULU. . . . "Midway," of course, referred to Midway Island.

Thus far, radio bulletins from the *Sea Dragon* indicated that she was encountering cooperative weather and clocking better than 100 miles a day. If that rate continued, her ETA in San Francisco would be along about the middle of May. Halliburton must have been fretting. They were about two months late in their timetable; and with the San Francisco exposition already open, there could be concern about whether or not space would be held for their vessel.

Much greater worries were coming over the horizon.

Trouble Brews

On March 21st the sky grew ominously overcast. The wind gained velocity, churning the sea into a vicious mood. And that was only a preview. Conditions worsened steadily the next two days. The wind built to gale force, howled even harder in gusts. Twenty- then thirty-foot waves assaulted the junk; rain squalls lashed her deck. Navigating was next to impossible.

On March 23rd the U.S. liner *President Coolidge* was knifing toward Yokohama, some 1,800 miles west of Honolulu. She picked up the *Sea Dragon*'s radio and sent greetings. The liner's chief officer suggested that the two vessels might rendezvous in a couple of days, conditions permitting.

But by now the junk was beating her way into a situation that threatened a full-blown typhoon. Even so, on the morning of the 23rd, her skipper transmitted a message that showed he hadn't lost his sense of humor completely: CAPTAIN JOHN WELCH OF THE SEA DRAGON TO LINER PRESIDENT COOLIDGE . . . SOUTHERLY GALES RAIN SQUALLS LEE RAIL UNDER WATER WET BUNKS HARDTACK BULLY BEEF HAVING WONDERFUL TIME WISH YOU WERE HERE INSTEAD OF ME.

The big liner was having problems of her own. She too was being slammed by steadily worsening weather and savage seas. Giant waves crashed over her bow, causing her to pitch and shudder alarmingly. Seasickness casualties ran high. Some passengers were bruised and otherwise injured as they were flung about and caromed off bulkheads like billiard balls. It boggles the imagination to try to picture what it must have been like on a seventy-five-foot junk.

Along about noon that day, Captain Welch fired off another

message to the *President Coolidge*: 1200 GCT 31.10 NORTH 155.00 EAST ALL WELL WHEN CLOSER MAY WE AVAIL OURSELVES OF YOUR DIRECTION FINDER REGARDS WELCH. That was to be the *Sea Dragon*'s last message.

Without stars or sun with which to take navigational fixes, Captain Welch had had to calculate his vessel's positions by dead reckoning; and under the circumstances this wasn't too certain. As of March 23rd he figured their location to be in an area approximately 900 miles southeast of Yokohama and 1,500 miles west of Midway Island. He and the liner's skipper thought they should pass within a few miles of each other during the next couple of days, at which rendezvous the *Sea Dragon* would be able to pinpoint her position.

But that wasn't to occur. A rendezvous now was impossible.

Deadly Silence

The liner tried repeatedly to contact the junk the rest of that day and on through the night. No luck. They hoped the silence was due only to storm damage to the *Sea Dragon*'s antenna, or perhaps a problem with her transmitter. When further calls went unanswered, the *President Coolidge* altered course somewhat and posted extra lookouts.

Next day, the storm's fury abated a bit, but the liner passed through the general area where the *Sea Dragon* should have been without sighting her or raising her on radio. A U.S. freighter, the *Jefferson Davis*, bound in the opposite direction, also passed within a few miles of where the junk should have been, but neither saw her nor heard any radio signals.

When the *President Coolidge* docked in Yokohama her captain and first officer were convinced that the *Sea Dragon* had gone to the bottom of the Pacific. For a while, however, hope was kept alive by frail speculations. It was noted that the junk had been in the horse latitudes, a globe-circling belt, roughly thirty degrees above the Equator and thirty below, in which there is often very calm, windless weather. Here, if she were conserving fuel or experiencing engine difficulties, she could drift for days or weeks, but without serious problems because she carried provisions and water for three months. In some quarters a chance remark by Halliburton to a friend before the voyage offered a measure of optimism. He had hinted that the *Sea Dragon* might just deliberately "disappear" temporarily to create extra suspense.

The U.S. Coast Guard in Hawaii pointed out that radio silence didn't necessarily mean conclusively that a vessel was lost. True, the little ship had been struck by a fearsome storm, but some maritime experts supported Halliburton's confidence in the extraordinary seaworthiness of junks. The *Sea Dragon* might well be continuing her course toward Midway Island.

Meanwhile, transpacific ships and aircraft were alerted to keep eyes open for the missing junk. Early in April the U.S. Navy detoured the cruiser *Astoria*, on her way home from a diplomatic mission to Japan. Turning back from Guam, the warship and her seaplanes searched thousands of square miles of ocean—in vain.

In the minds of some experienced observers there was no doubt that the *Sea Dragon* was gone. One of them—veteran skipper Charles Jokstad of the liner *President Pierce*—had inspected her at Halliburton's request at the yard in Hong Kong, after which he emphatically warned the headstrong adventurer against the voyage. In Captain Jokstad's opinion, heavy seas amputated the junk's rudder and rendered her helpless in a broaching-to situation—that is, wallowing in a trough and turned broadside to oncoming mountainous waves. After that it was just a question of very little time. She probably was dismasted and broke up quickly, he added.

Requiem

Word of Halliburton's disappearance flashed all over the world. Thousands of letters cascaded into newspapers, his publishing house, and his parents' home. Such was the man's reputation for conquest, many were sure that he survived on some remote, uncharted Pacific isle, and would be rescued to write about his greatest escapade of all.

As weeks became months, however, even the most sanguine of his fans lost hope. The final curtain rang down on Richard Halliburton when he was declared legally dead.

Wherever Richard Halliburton is, we can speculate that the failure of his last venture probably bothers him more than having died. One thing is certain: He would have found a large measure of satisfaction in the tens of thousands of readers who remained loyal to the end—and beyond.

8

The Calamity Line

Fifteen years after the proud steam-and-sail liner *President* departed New York for Liverpool on a west-to-east transatlantic voyage that carried her passengers and crew to oblivion without a trace, a passenger ship named *Pacific* duplicated the feat of maritime mystery during an east-to-west crossing from Liverpool to New York.

Smaller than the British *President*, the *Pacific* operated in a fleet owned by the United States Mail Steamship Company, more commonly known as the Collins Line, after its founder, Edward K. Collins. It cost anywhere from about $70 to $150 (first class) to cross the Atlantic in those days, and the *Pacific* got her share of passenger revenue.

On the twenty-third day of January, 1855, she sailed from the Mersey River estuary, Liverpool, outward bound for her home country. Because of the Atlantic Ocean's reputation for unsociable moods in midwinter, the passenger list was small, only forty-five persons on this particular trip. They were outnumbered by a crew of 141. In her belly was a good payload of cargo.

The *Pacific* never reached New York. Somewhere out on the ocean's lonely expanse she vanished completely. No trace of her—wreckage, survivors or victims—was ever found. Lacking even a shred of evidence, the Collins Line could only speculate that she had a fatal encounter with an iceberg. That may very well have been the case; but, like her equally unfortunate predecessor, the *President*, she joined the sea's collection of unsolved riddles.

Loss of the *Pacific* was only one in a series of tragic chapters in the history of the Collins Line.

Twists of Luck

Right from his entry into the ocean shipping business Edward Knight Collins earned the respect of rivals. He began with notable success, operating a fleet of five handsome clippers in the transatlantic trade. His *Garrick, Roscius, Shakespeare, Sheridan* and *Siddons*, so named for prominent figures in the theater, averaged only about 900 tons apiece, but they funneled money into their owners' coffers as they they sailed back and forth between the United States and England in the 1840s. Also about this time, a man named Samuel Cunard and his English associates formed the British and North American Royal Mail Steam Packet Company (this became the world-renowned Cunard Line in 1878). An intense rivalry developed between the American and British shipping firms.

Up until then Mr. Collins and associates had been satisfied with their canvas-powered clippers—and why not, they were lucrative. But when Mr. Cunard and associates introduced four new steamships in the competition, his American rival couldn't help but notice their immediate success. The Collins clippers were fast; under ideal conditions they could even beat some of the new Cunard steamers for a distance. But the steamships were faster, overall, and, what was more important, they were more consistent in that speed. At their very best, clippers might make an east-bound crossing in an exceptional time of seventeen days. With the whim of wind and weather, however, an average for sailing vessels was more like twenty-two days east-bound and thirty-four west-bound (the Gulf Stream's eastward sweep aided the former, but had to be bucked in the latter). In contrast, Cunard's steamships could uniformly average about fifteen days east-bound and seventeen days in the opposite direction.

Edward K. Collins had been following the development of steam propulsion with lively interest, and the success of the new Cunard vessels convinced him that the days of sailing ships were numbered. Translating his conviction into action, he formed the United States Mail Steamship Company* and, aided substantially by a subsidy from the U.S. Government, ordered construction of four steam-powered vessels: *Arctic, Atlantic, Baltic* and *Pacific.* Built from wood, strongly reinforced by iron rods, they averaged approximately 2,600 to 2,800 tons each (a far cry from the Atlantic's later 50,000- to 80,000-ton liners) and were propelled by two beam engines turning large paddlewheels, one on each side. The 2,794-ton *Arctic,* for example, had twin engines of about 1,000 horsepower each, swinging 25 1/2-foot paddlewheels.

Vital in competition for passenger patronage and freight contracts, these new Collins entries were faster than some of Samuel Cunard's steamers and provided more sumptuous accommoda-

*The word "Mail" appears in the names of both lines, Collins and Cunard, for a good reason. Transporting mail was very worthwhile, since it brought financial support in the form of subsidies from governments.

tions. With these advantages they became very popular and soon were making a large dent in their British rivals' income.

Unhappily, though, the fortunes of the Collins Line were about to take a bad turn. There developed a wide disparity in the luck of the two lines. Collins, it turned out, was to be singularly unfortunate, whereas it was very much the other way around for Samuel Cunard.

Tragedy Hits Home

Double misfortune overtook Edward K. Collins and his line in 1854. For him it was the most shattering blow of his life.

In August of that year he bid bon voyage to his wife, his daughter Mary Ann, and son Henry, off on an eagerly awaited holiday trip to England aboard his line's *Arctic*. In September, Collins' family was among some 281 passengers (figures vary) on board for that steamer's return crossing. On September 20th they sailed from Liverpool, the ship commanded by veteran master James C. Luce. Also on board was Captain Luce's eleven-year-old son Willy. The lad was crippled, and his parents thought a sea voyage would do him some good.

Catastrophe struck when the ship was only forty to fifty miles off Cape Race, Newfoundland.

In dense fog, a frequent navigational hazard in that region, the *Arctic* collided with an outbound French steamer, the *Vesta*, commanded by Captain Alphonse Duquesne. In an ironic twist of fate, Captain Luce made a humane but fatal miscalculation. Thinking the much smaller *Vesta* was mortally wounded and in imminent danger of sinking, he dispatched one of his boats to see if the other vessel required assistance. What he didn't know was that the *Vesta* had a steel-plated, compartmented hull. The French steamer was, in fact, in better shape than the *Arctic*, despite the appearance of her bow and the panic of the people on her deck.

Captain Luce quickly discovered that his own vessel was injured seriously. By tons, the Atlantic Ocean was invading her noncompartmented hull through gaping wounds. Efforts to stem it with canvas failed. Although he wanted to aid those aboard the seemingly doomed *Vesta*, Captain Luce's first responsibility was his own stricken ship. He ordered her to proceed to the nearest landfall with all possible speed. In a side-calamity, a *Vesta* lifeboat filled with passengers who had abandoned ship against Captain Duquesne's orders was run down in the fog and ground to kindling by the *Arctic*'s starboard paddlewheel as she steamed away.

The *Arctic* lost her race for safety. Water poured into her in such great volume that her pumps couldn't handle it, and when it reached her furnaces she was finished. Only twenty miles from the nearest coast she started to go down by the bow.

Nightmare

Survivors later described the pandemonium of the steamer's final hours.

The *Arctic* satisfied contemporary maritime laws as regards her lifeboats and other lifesaving gear; but, incredibly, those regulations did not specify adequate equipment. There were not enough lifeboats to go around. Survivors told about panicked engine room men and other crew members, along with a number of male passengers, defying Captain Luce and his officers as they forcefully shoved other people aside and battled each other to crowd into the lifeboats and launch them. It was reported that an officer who tried to stop them was killed by a stoker. No "Women and children first!" in that insane melee. The fierce scramble for lifeboats even interfered with attempts by other crewmen and male passengers to fashion a large emergency raft. To make matters worse, some life preservers were unusable.

But there were acts of courage and heroism too, by crew and passengers. Slowly the *Arctic* settled by the bow, her stern starting to lift for her plunge into the deep. Emotions ran a full range among those still aboard. Some became hysterical; others, helpless, wept piteously; a few joined in singing hymns; still others waited silently in stoic calm. An untold number fell or were swept into the sea as the angle of the ship's deck increased steeply. Poised for a moment with her stern ever higher, she slid beneath the waves.

At least one lifeboat filled with passengers had managed to get away. A few managed to hastily put together emergency rafts. Many other passengers and crewmen were in the sea and clinging desperately to anything that floated. For them death was as certain as by drowning. They were in paralyzingly frigid water, where survival is measured only in minutes. Somewhere a brand of justice caught up with a lifeboat selfishly commandeered by a group of crewmen. They were never seen again.

Under the circumstances, it borders on the amazing that anyone survived. During days following the disaster, vessels that included the *Cambria,* the *Lebanon,* an unidentified whaler, and several Newfoundland fishing boats picked up the lucky ones and drifting bodies. Tallies of survivors and known dead or missing have varied, but these figures are believed to be fairly close: From a list of 281 passengers, 23 were saved, along with a man from the *Vesta* (about 8.5%). From a list of 153 officers and crewmen, there were 61 survivors (approximately 40%).

Among those lost were the wife, daughter and son of Edward K. Collins. In a tradition of the sea, Captain Luce went down with his ship, and by his side was Willie. Both were sucked under by a swirling vortex as the *Arctic* headed for the bottom, but both miraculously popped to the surface. Sadly though, the miracle was not complete. Before his father could get to Willie, a paddlewheel housing was torn free from the sinking ship and was literally shot to the surface by air trapped inside. As it broke water it rolled over on the lad, killing him instantly. Ironically, this same huge piece of superstructure was instrumental in saving Captain Luce's life and he eventually was rescued.

The *Vesta* not only survived the collision, but also limped safely into port. In still another touch of irony, had the *Arctic* remained on the crash scene, the French steamer probably could have saved all on board.

All Downhill

Such was the route of the Collins Line's fortunes. It survived loss of the *Pacific*, but after the *Arctic* disaster the company became increasingly enmeshed in financial difficulties. As things worked out, however, it was politics, not sinkings, that delivered the *coup de grâce*.

Although the U.S. Civil War had not yet erupted, the slavery issue was starting to glow an angry red and drive a wedge between North and South. Collins' vital federal government subsidy for carrying mail became a casualty in Congress when a Southern faction successfully argued that the line benefited only the North. Without this margin of government financial support, the Collins Line had no hope of competing with subsidized British rivals. It folded, and creditors forced sale of its ships.

In a final bit of irony, the very factors that had made the Collins Line so successful earlier—speed and elegant accommodations— contributed to its ultimate downfall. Being faster and fancier than English competitors, the Collins steamships cost more to build and operate. This led to charges of "gross extravagances" with the public's money, since a government subsidy was involved. Those charges, in turn, provided powerful ammunition with which its opponents could "torpedo" the line.

9

The Day the Indian Ocean Exploded

For decades a miles-long fracture in tables of bedrock underlying the U.S. Pacific Coast has been giving Californians an understandable case of jitters. This fracture, which geologists labeled the San Andreas Fault, represents a kind of disagreement between two fantastically enormous masses of rock, with one determined to shift northward against resistance provided by the other along the Fault. So long as the shifting mass can creep, however slowly, there's no great threat. But if the resistance of its opposing mass is such that it can't move at all, an incredible pressure builds until the northerly-shifting mass of rock finally has its way and, out of sheer force, suddenly breaks free. When that happens, you have shattering results on the surface. Specifically, you have an earthquake.

Such was the San Andreas Fault situation that caused the catastrophic San Francisco earthquake of 1906. With the holocaust and devastation that followed, it was one of the greatest disasters of modern times. Among the grim chronicles which history keeps of such doomsday affairs, San Francisco's 1906 earthquake might well stand with such spectacular calamities as the eruption of Mount Vesuvius in 79 A.D., which leveled the ancient city of Pompeii and barbecued all its inhabitants.

As terrible as those catastrophes were, however, neither matched the self-destruction of the volcanic island of Krakatoa (or Krakatau) in 1883 for sheer explosive violence. Krakatoa's cataclysmic suicide, and the fearsome tidal wave which followed, had effects which were felt around the entire world.

Where Islands Go
Up and Down Like Elevators

As on land, there has been troubled unrest beneath ocean floors in many parts of the world since our planet was formed. One kind of trouble has been caused by great faults or fractures in the globe's crust; the shifting of vast tables gives rise to marine earthquakes, commonly called seaquakes.

Over millions of years there has also been undersea volcanic activity, thrusting ocean floors skyward to break the surface. Numerous archipelagos are in reality the exposed tops of submarine mountain ranges. The State of Hawaii is a case in point. Some oceanic islands are the summits of mountains which, if measured from their bases at the bottom of the sea, are higher than 29,000-foot Everest, the mightiest land mountain. When so measured from the Pacific's bottom, 18,000 feet down, and adding its 13,784-foot exposed portion, Hawaii's extinct volcano Mauna Kea becomes the world's tallest mountain, roughly 32,000 feet high.

Numerous oceanic islands, in the Atlantic and Caribbean Seas as well as the Pacific, are the products of volcanic upheavals. Many have long since retired from active-volcano activity. Others, like Hawaii's 13,680-foot Mauna Loa, and the volcano La Sourfriere in the French Antilles in the Caribbean, continue to mutter threats from time to time, their fiery stomachs belching acrid smoke and vomiting molten lava. In remote regions of the Pacific Ocean, undersea volcanic activities cause new islands to appear, old ones to disappear, and some to alternately appear, vanish, and reappear.

Krakatoa was the offspring of such activity.

It was a volcanic island which poked through the surface of Sunda Strait, between Java and Sumatra, way back in a remote past. Krakatoa was more a volcano than an island. Indeed, it was a large volcano squatting on the strait's floor. What we might call Krakatoa Island was its pinnacle, a crater which had thrust itself far enough upward to appear as an island in the East Indies seas.

No one knows for certain when Krakatoa came into being. A geological theory once proposed that much of what is now Sunda (also Soenda) Strait is the excavation left by a supervolcano that blasted itself to kingdom come eons before the first humanoid creature trod the earth. A speculation is that this supervolcano blew its top, leaving only shattered fragments of its gigantic base, appearing as a ring of small islands. Krakatoa was an isle in that ring. Through subsequent developments deep under the ocean bottom, Krakatoa became an active volcano in its own right, picking up where its parent left off.

An Early Message

The first broad hint in relatively modern times that Krakatoa might have a sour disposition came in 1680. In the spring of that year the volcano erupted. Huge clouds of sulphurous smoke rose high above the island, and from its cone flowed orange-red rivers of melted rock. Having thus vented its spleen, Krakatoa simmered down. Except for periodic rumbles and mutterings, accompanied

71

by occasional belches of smoke and hisses of steam, Krakatoa dozed and remained reasonably quiet for two centuries.

The next warning that the brooding volcano was up to some deviltry came in the 1880s. The prelude was a series of seaquakes. Then, in May of 1883, a widening cloud of steam was observed hovering over the crater, and fissures about the cone poured ominous-looking smoke into the air. Krakatoa's grumbling, punctuated by hisses of steam, grew louder. It soon became obvious that the volcanic island was readying some surprises—all unpleasant. The shattering culmination that was to occur in a few weeks undoubtedly had its beginning in volcanic activity taking place on or about May 20th that year. But even with ominous warnings no one was prepared for the big blowout that began on August 26, 1883.

The Bomb Ticks

By August, crevices split Krakatoa open, and the sea rushed into the breaches. Next came a reaction that set the stage for an extravaganza without equal, before or since. When sea water rushed into the crevices it cooled the molten rock inside the volcano, temporarily sealing off its throat. Uncountable millions of gallons of water were now trapped inside Krakatoa's fiery abdomen and converted to steam, building to a pressure of incalculable magnitude. Krakatoa became a gargantuan time bomb. Its fuse was set, so to speak, and at any moment it would explode.

The moment of detonation came on the afternoon of August 26th. The stupendous fireworks began with a series of preliminary salvos as some of the weaker fissures let go. These introductory explosions were insufficient as a safety valve, and the next day they culminated in a blast to end all blasts. In this detonation, mightiest ever known to man, Krakatoa blew itself into oblivion. In a shattering red and white flash, two-thirds of the island vanished. Where only seconds before a mountain had risen 1,600 feet toward a tropical sky, there now was a vast hole 1,000 feet deep beneath the waves. Krakatoa had torn out its own guts, pulverized them, and blown them to the winds.

Nothing, natural or man-made, in recorded history has equaled the noise generated by Krakatoa in its death throes. The blast was heard in Australia. It also was heard in Manila, nearly 2,000 miles away, where residents thought a ship had fired distress signals and made ready to send a rescue vessel. It reportedly was heard in Madagascar, roughly 3,000 miles distant, which is comparable with people in New York hearing an explosion in Los Angeles. Shock waves set in motion by the eruption registered on barographs completely around the world.

Aftermath

The most terrifying effect of Krakatoa's violent suicide was a tidal wave of unbelievable size and velocity. Right after the blast a towering wall of water at least 100 feet high hurled itself toward the coasts of Java, Sumatra and adjacent islands at speeds variously estimated at between 400 and 700 miles an hour. At such velocity, the islands had little or no warning as the monstrous mountain of water hurtled toward them. Reaching those coasts, its mass and momentum carried it inland. Across beaches and over lowlands it roared, still fifty feet or more high, engulfing everything in its path. People, animals, houses—entire villages—disappeared in the savage flood. In a fit of whimsy the tidal wave plucked a Dutch gunboat from her moorings and carried her two miles inland, leaving her stranded thirty-five feet above the coast's high-tide level.

It was calculated that 36,500 people were killed by the tidal wave on Sumatra and Java; and that could have been a conservative estimate, increased by a toll on other islands. The numbers of domestic and wild animals destroyed were beyond calculation.

The tidal wave generated by Krakatoa's death spasm sped out across the Pacific and was still evident by a seven-inch height by the time it rolled into San Francisco Bay, thousands of miles away. Remnants of the wave lapped as far as Indian Ocean shores, and even beaches at South America's Cape Horn felt the effects. Like a windjammer in the old days, it rounded the Horn and, still traveling at a respectable rate, moved into the Atlantic Ocean, where it reached as far north as the English Channel, around on the other side of the globe.

In its self-annihilation Krakatoa blew pulverized rock as high as twenty miles into the sky. Volcanic ash and pumice reached the stratosphere, and at that level were carried by winds three times around the globe. The finest dust remained suspended in the upper levels for months and for about a year afterward caused spectacular sunsets in every country in the world. Some places reported the air-borne layer of fine volcanic dust was intense enough to create an eclipse effect, necessitating the use of lamps by day. And in 1884, one year after the blast, pyrheliometric observations (measurements of how much of the sun's heat reaches the earth's surface) revealed that the sun's warmth was still thirteen percent below normal for that time of year.

The volcanic by-products of Krakatoa's disintegration reached a volume boggling the imagination. A contemporary account told of ashes being distributed over an area of 300,000 square miles. The greater part of this debris fell within an eight-mile radius of the former volcano. Said one account at the time: "Stretches of water that had an average depth of 117 feet were so filled up as to be no longer navigable. Enormous masses of pumice floated upon the sea and stopped navigation except for the most powerful steamers." Nature, it seems, can be a worse water-polluter than man at times.

Post Mortem

Such was the violent valedictory of Krakatoa. Most of the volcanic island had vanished. All that remained was a forlorn, jagged fragment, a gray and desolate islet.

Sometime afterward, when it seemed that Krakatoa was truly dead, a handful of scientists landed to examine what was left of the body. Needless to say, they didn't expect anything alive in human, animal or vegetable form. Any references to natives on the island prior to the blast are very vague at best. Only one account made fleeting mention of "inhabitants." Fact is, there was no way of knowing whether or not anyone resided there. This is for certain: If the island had tenants, they departed swiftly and in all directions when the big boom came. Presumably, any people on the island left when warnings grew more intense. If any waited until the last minute before the blast, they most certainly were victims of the nightmarish tidal wave.

It's not entirely accurate to say that absolutely no trace of life was found on Krakatoa after the eruption. Scientists examining the shattered remnants found one tenant. A solitary, tiny spider was busily constructing an equally tiny web in the best "life must continue" tradition. It unquestionably had been brought there by the wind.

Krakatoa's sound and fury have long since subsided, and the four winds have dispersed the last of its dust. But there's an epilogue.

In 1929, in the area so violently departed by the volcano, there arose from the sea a new island, pushed upward by those restless stirrings beneath the ocean floor. To this sea-spawned island they gave the name Anak Krakatoa, Child of Krakatoa.

10

Biggest Jinx of All Time?

Not far off Montauk Lighthouse, at the easternmost tip of Long Island, a mass of stone squats on the Atlantic's floor and thrusts a menacing spire to within twenty-four feet of the surface at mean low tide. Fishermen and mariners know this flinty steeple as Great Eastern Rock, and as such it can be seen on navigation charts of the area.

The tower of stone was named for the steamship *Great Eastern*, largest vessel of all in the 1800s and certainly one of the unluckiest vessels in maritime history. A permanent Jonah* was created in the form of the *Great Eastern* when she was built, and her story blends pathos, comedy, and dashes of tragedy, then concludes with exposure of a grisly secret that adds an eerie, haunting finale.

Her mystery is that she lived as long as she did.

From the Very Start, Bad News

The *Great Eastern* emerged from the brain of a creative genius with an unusual name, Isambard Kingdom Brunel, one of the most brilliant engineers of his era. He designed and built railroads, steamships, bridges, and other great engineering projects.

From Brunel's blueprints the huge steamship took form at Isle Dogs in the Thames River, London. With a hull 693 feet long and 120 feet in breadth, and 22,500 tons displacement, she would be the mightiest ship for nearly fifty years. Furthermore, with engines developing a then-astounding total of about 11,000 horsepower, harnessed to a pair of large paddlewheels *and* a 24-foot propeller, supplemented by 58,500 square feet of sail, she also would be the fastest ship to come over the horizon thus far. In addition to transporting big-revenue freight cargos, Brunel visualized her as accommodating up to 4,000 passengers—an over-optimistic figure, it turned out.

---

*Despite the ultimate good fortune of the Biblical prophet Jonah, who was cast into the sea during a storm, swallowed by a whale and later cast up safely on land, among seafaring folk his name was given to anyone whose presence was supposed to bring bad luck.

The *Great Eastern* was revolutionary and ahead of her time in still other ways. Her hull was double—one shell inside another, with three feet of space between them: a safety feature if the outer hull were punctured. There were no ribs as such, but bulkheads strengthened her hull and provided the safety feature of watertight compartments. Like another leviathan that followed many years later, the *Titanic*, this giant was described as unsinkable by her architect. In both cases the validity of the claim was to be challenged, but with radically different outcomes. Whereas the *Titanic* would have her unsinkability violently disproved by an iceberg which disemboweled her and sent her to the bottom with an appalling loss of life (on her maiden voyage, no less), the *Great Eastern* would survive an encounter with one of King Neptune's assassins and eventually perish in pedestrian style, but without demanding that any of her guests share her demise.

Thousands of steel plates, held together by millions of rivets, went into fabrication of the *Great Eastern*. Some 2,000 workmen labored to put her together. Even with this work force, construction was long and drawn out, due chiefly to financial problems. What with one thing and another, including bankruptcy of her first company, nearly six years elapsed between laying-down of her keel and complete outfitting.

As in any construction project of such size, a certain casualty rate among workers was anticipated, but there were times when the *Great Eastern* seemed to be trying to overdo it. One worker fell to his death. Another died between the hulls. One of the boy helpers took a header from aloft and was skewered on an upright metal rod. Another workman toppled off a scaffolding and fell on a cohort, literally braining him. Even ''sidewalk superintendents'' got into the grim act. An onlooker, inspecting a piledriver too closely, had its massive ram drop on his head. And there was a note of mystery. One workman abruptly disappeared from his job and home, never to be heard from again.

At Long Last

The giant finally was ready for launching. This may have been the day her Jonah went on permanent active duty. Fraught with all sorts of discouragements, her debut was hardly what you'd call a festive occasion.

To begin with, they couldn't get her big, fat behind off the riverbank and into the Thames. Launchers tried to shove her sideways, but she wouldn't budge. Hydraulic rams were trundled in as persuaders; and a great chain, spooled on a huge, man-powered windlass, was secured to the ship to prevent her from leaving the party altogether.

The hydraulic rams groaned. The *Great Eastern* started to slide toward the water fitfully. Suddenly there was a lurch that brought the checking chain up tight on the windlass, setting its drum to spinning like a Ferris wheel gone berserk. Its heavy wooden bars, now whirling, caught a handful of workers and flung them at the crowd. The spectators could take a hint; they ran for home and mother. So did many of the shipyard workers. The sight and sound of the steel behemoth sliding sideways toward the Thames so unnerved the crews of barges standing by to assist that they dove overboard. Minutes later the *Great Eastern* ground to an adamant halt, and the first attempt at launching ended with her being stuck on the ways in a torrential downpour.

Someone said that Brunel himself ordered the first launching halted, suddenly fearing that the resultant waves might drown half the spectators lining the Thames shores. In any case, the ship was jammed securely on her launching ramp. There followed a long and costly delay before she finally was floated, an unforeseen setback that contributed to the bankruptcy of her first owners.

Undaunted, Brunel organized a new company and bought the ship. But the earlier launching difficulties, bankruptcy, and the strain of refinancing left him a sick man. On a visit to his brainchild under her new ownership he was felled by a stroke and died not long afterward. The day before he died, one account has it, there was some kind of explosion aboard the ship that killed several men.

But these grim notes were forgotten temporarily as the *Great Eastern* basked in glory as the wonder of her day. Here was an incredible marvel. As if her size alone were not enough to make beholders' jaws dangle, she carried six masts and five funnels, a double record in itself. She was so impressive that she caused other vessels to pile into docks and each other when they saw her for the first time. When the *Great Eastern* was opened to visitors in New York on her first voyage to the United States in 1860, nearly 150,000 people came from far and wide to pay for the privilege of gawking at her saloons and staterooms. Two great poets, Whitman and Longfellow, were inspired to wax lyrical about the luxury vessel, and the famous art firm of Currier & Ives was sufficiently stimulated to create dramatic lithographs of her to decorate parlors across the nation. One of the few discordant notes came from author Herman Melville (*Moby Dick*). He pronounced her a "vast toy," bleakly predicting that she "can't last a hundred years hence." Old Herman wanted his money's worth.

Seafarers in the old days were a superstitious bunch. Many of their superstitions sounded ridiculous, yet in the young *Great Eastern* they already had a showcase of "proof." The delays and fatalities attendant upon her construction, bankruptcy of her owners, and premature death of Isambard Kingdom Brunel spelled one thing loud and clear to the old-timers: This new marvel was a hoodoo, and a dangerous one at that. Subsequent episodes in the liner's life did nothing to contradict them.

During a trial run, a funnel blew up like a giant firecracker, fracturing all the mirrors in the grand saloon and rearranging several staircases. Immediately following the blast, people aboard were kept busy dodging a shower of debris that included shards of glass and metal fragments. Five members of the engine room's "black gang" were fatally burned and others were injured. Another crewman, fleeting the explosion, fell into a paddlewheel and was killed. A large fortune already had been poured into the *Great Eastern*'s furnishings. Now came expensive repairs and refurnishing.

On another trip, the dinosaur of a steamer sashayed into a pier, and a paddlewheel chewed away a large portion of it before things were brought under control. Periodically the *Great Eastern* took out a grudge she seemed to have against piers. These mishaps caused a newspaper wit to suggest that the giant could save engineers a lot of money in digging the Panama Canal by ramming her way across the isthmus.

There was no end to her bad luck.

During a particularly violent autumn storm the big steamship tore free of her moorings and was driven out into open ocean. She rode out the storm all right, but wind and water damage necessitated additional costly repairs. A couple of months later, her first skipper, Captain William Harrison, and two others from the ship were drowned when a sudden squall capsized their smallboat during a trip to shore. This tragedy seems to have precipitated organization of still another owning firm.

Perhaps it's just as well Isambard Kingdom Brunel couldn't know about his pride-and-joy's maiden voyage to New York in the spring of 1860. Delays in making ready cost her an untold number of cancellations; and when she did finally sail, a month late, there were less than 100 paying passengers aboard, a far cry from the 4,000 Brunel had envisioned. And they had an awkward time of it for a while when problems with the funnels caused evacuation of the dining saloon.

Her arrival in New York brought all the traditional hoopla accorded a new liner making her debut in that port: Crowds of rubbernecks on shore, a large escorting fleet of miscellaneous vessels, and incessant tooting of whistles. But the *Great Eastern*'s Jonah was unimpressed by the hospitality. On arrival at her designated pier, one of her paddlewheels again chewed away a generous portion of it. And that was only for openers. While making ready for visitors, a crewman was killed when he fell into a sidewheel, and another fell overboard and was drowned. The crew did not get along well together apparently. There were frequent brannigans, and when fights killed two and injured more than a dozen others, police were brought aboard to keep the peace. When visitors finally got aboard, they showed their resentment at the one-dollar admission price by trying to steal everything that wasn't nailed down.

Nor was that all.

During her New York stay the *Great Eastern* offered a special two-day excursion, like the "cruises to nowhere" that came many years later. Bonus features of this delightful trip included: Passengers actively ill from either *mal de mer* or drinking too much, or both; card games marked by fights; food ruined by a defect in the provision compartment; bedding arrangements for only 300 guests, forcing some 1,700 to sleep on deck, where they alternately were festooned with soot and pelted by rain. Understandably, the excursionists could hardly wait to leave this bad dream at the ship's scheduled port of call, Old Point Comfort, Virginia. But even this didn't go right. For some unexplainable reason, the *Great Eastern* was far off course. The cruise's one happy feature was that the ship carried lots of liquor. Torn between anger and concern at the delay in reaching Old Point Comfort, the passengers mass-assaulted the liquor supply and on empty stomachs got drunk as never before. Needless to add, they ended their pleasure cruise the minute a dock was available. It's said that the excursion's return trip to New York was enjoyed only by stowaways, and that those who could be rounded up were charged fifty cents for the voyage.

The *Great Eastern* fared little better on her return crossing from New York. There were less than 100 paying passengers aboard. But at least she satisfied Brunel's predictions concerning her speed. She made the east-bound crossing in nine days and four hours, a new record. In two mishaps a few months later she collided with a frigate and tangled with a small boat, drowning two of the latter's passengers.

The huge steamship never did prove profitable, passengerwise or in cargo revenues. Isambard Kingdom Brunel had visualized her in the Australia and Far East trade, where she might have done well. In transatlantic service she just didn't pay. Reports in the press of her misadventures didn't help, of course. Even the British Government

79

wasn't interested in taking this floating calamity off her owners' hands.

A Near Thing

Nature very nearly relieved her weary owners of the *Great Eastern* by smacking her with a particularly fierce storm during a crossing in the autumn of 1861. The resulting pandemonium had both near-tragic and serio-comic aspects.

Most frightening was a concern that the big steamer would break in half. No ship of this size and type of construction had ever battled such a storm, so there was no precedent by which to judge. There was agonizing suspense, but no breaking into halves. A more tangible danger was the damage to her steering apparatus. Being out of control for long in those conditions could have been fatal. For once the *Great Eastern* was lucky. Her crew was able to jury-rig emergency steering to hold course through the storm.

Meanwhile, the grand saloon was a scene of low comedy. An enormous grand piano waltzed gaily around the big room, skating from one side to the other, narrowly missing scampering passengers. Finally it fetched up against a bulkhead with a mighty crash and a wild chord, amid a jumble of splintered wood and tangled piano wires. Chairs skittered and danced across the saloon. Big sofas crossed and recrossed the room at breakneck speed, chasing people running for the nearest exits. A large stove decided to join the party and broke loose. In the melee it knocked a passenger galley-west, cutting his head, knocking out a tooth, and dislocating a finger.

Livestock was still carried aboard for fresh meat in those days. Outside on deck, breaking waves demolished pens and cages, liberating a lively assortment of animals and fowl. A swan found his way into the grand saloon, and there tried to make the most of this indoor aerodrome by taxiing back and forth on its polished floor in efforts to get up enough speed for a takeoff. A huge wave flung two cows into a ladies' room, surprising all concerned. And in the midst of all this bedlam, tons of water entered the ship through broken windows and other openings and cascaded down through the decks. In all, it could have been a scene from Noah's Ark.

In a single day the ship's surgeon administered to twenty-seven major fracture cases, plus a large assortment of injuries that included broken noses, multiple contusions, lacerations, and purple mice about the eyes. At one stage there was even a semblance of a mutiny. While the ship was temporarily disabled, with her engines stopped, some of the engine room crew refused to obey orders and broke into the liquor stores. Fearful of violence, the captain appointed a vigilante patrol of male passengers and issued arms to them. This situation seems to have been of short duration, however.

Jonah and the Rock

The calendar advances to August 1862. In the United States, any reported escapades of the *Great Eastern*, if noticed at all, would have been submerged in news of the Civil War, now a bloody reality.

That month the white elephant was on her way to New York, on what may have been one of her relatively few profitable crossings. Her holds were pregnant with freight. More than 800 passengers relaxed in their cabins or lolled in the liner's fancy public rooms, her past misadventures forgotten as they enjoyed travel aboard such a modern, posh, floating hotel. Three thousand miles away, roughly east by north of Montauk Lighthouse, a stony steeple in the sea was waiting to write another chapter in the *Great Eastern* saga.

This crossing was uneventful. A week and a half after leaving Liverpool, the steel giant cruised into the offing of Long Island on a glassy, moonlit sea and soon drew within sight of Montauk Point. Under the benevolent Cyclops eye of its venerable lighthouse the *Great Eastern* prepared to enter Long Island Sound. Because of the vessel's considerable draft, her captain elected to approach New York via the deep-water sound, rather than risk treachery by shifting sand bars lurking on Long Island's southern coast. As his charge glided quietly through the moonlight of that summer night, he ordered the engines to proceed at slow ahead as he impatiently awaited a rendezvous with a pilot boat. A pilot would guide them through a then-erratic channel lying between Endeavor Shoals, just to the northwest, and the Montauk peninsula, thence into the safety of Long Island Sound's deep water.

Sometime after 0100 hours the *Great Eastern* had her rendezvous with the pilot's gig and resumed normal cruising speed on entering a channel south of Endeavor Shoals. A half-hour later, the pilot and others on the bridge heard a muffled rumble emanating from somewhere down in the water's dark depths. Simultaneously, the *Great Eastern* listed noticeably to port in protest, then righted herself without so much as a pause and continued as though nothing had happened. The disturbance was momentary, not even rude enough to stir sleeping passengers.

Captain and pilot were mystified by the incident, but the ship seemed none the worse for her experience, whatever it was. The pilot's speculation that she may have rubbed her belly against an uncharted sand bar allayed the skipper's worries for the time being. Neverthless, he sent a damage-control party below to search for any

81

signs of leaks. None were discovered, and the *Great Eastern* negotiated the length of Long Island Sound without further incident, fetching the bay at Flushing just as a new day's sun was peeping above the eastern horizon. There she made a scheduled anchorage, and in due time passengers debarked for a ferry to Manhattan.

Now the steamship's commanding officer could no longer dismiss from his mind the subtle occurrence of the night before off Montauk. The ship had developed a starboard list—not bad, but noticeable. He engaged a diver to descend and have a look at her hull below its waterline. Back on deck, the diver reported that the *Great Eastern*'s outer skin had been rent by a terrible gash estimated at more than eighty-five feet long and better than ten feet across. It was as though she had been cut by a gigantic can opener. Her double hull had saved her, conceivably preventing a real disaster. For a change, the steamer's luck was good.

Obviously, no shifting sand bar had inflicted this damage. And since icebergs had to be ruled out, the villain had to be a rocky mass of some sort. Subsequent hydrographic probing of the area where the *Great Eastern* sustained her wound revealed a huge rock formation skulking in the depths and thrusting a slender spire to within less than twenty-five feet of the surface at low water. Presence of this kingsize spike had not been suspected because of its depth below the surface. It remained for the *Great Eastern*'s draft to find it and—of course—scrape against it.

It was an uncomfortably close brush with an invitation from Davy Jones, but the steamer had survived. Repairs were effected right there in Flushing Bay, employing a unique cofferdam. Her wound closed, the great ship continued her way in a career notable more for downs than ups.

Kaleidoscope

In 1864 the *Great Eastern* acquired yet another owner and began a new career. At what may—or may not—have been a bargain price, she was bought at auction and became a cable layer. In the first phase of her new life she started out fine, gradually extending a transatlantic cable between Ireland and Newfoundland, then failed when she lost hold of the end of it, leaving behind more than 1,000 miles of cable. In 1866 she acquired still another owner, and this time succeeded in a transoceanic cable-laying project. Subsequently she added to her true accomplishments by laying additional cables and repairing others. In between she kept her hand in the hard-luck business for an interval during which she returned to passenger-freight service long enough to bankrupt a French company.

Otherwise, the ensuing years didn't treat her too kindly. The sunset of her life had humorous touches. One in a procession of owners had a brainstorm to use her in an immense advertising program. She came into Liverpool harbor—embarrassed, no doubt—carrying on each side, in letters thirty feet high, the advice that LADIES SHOULD VISIT LEWIS'S BON MARCHE CHURCH STREET. Then there was a pleasure cruise, also part of an advertising stunt, to which she retaliated with a boiler-room fire, bursting pipes, and engines that fiendishly quit in the worst possible locations. Among the *Great Eastern*'s last public appearances was one as a floating carnival. Tied up to a pier, she featured sideshows, dancehalls, refreshment concessions and all the whoopdedoo of a circus midway. Finally she was retired altogether, relegated to a backwater harbor, where her massive hulk was abandoned to slow disintegration by rust.

Sundown and Night

It was merciful, perhaps, that the once impressive queen of the sea was condemned to execution by wreckers before being humiliated further. At least it was a mercy to any additional owners that might have come along. She had been a burden and headache to most of those who had possessed her. During her existence the *Great Eastern* bankrupted seven companies, was involved in seven major collisions with other vessels, plus lesser encounters, plus destructive arguments with piers; consumed fortunes in repairs and reconditioning, to say nothing about operation costs; in one way or another was responsible for the loss of twenty or more lives, and injuries to numbers of people; and, it was said, had experienced five mutinies or mutinous acts by her crews. Few other ships have known so many different owners and masters, or made them vacillate so much between pride and serious thoughts of blowing their brains out.

The *Great Eastern* had a long, if not good, life. Thirty-one years, pretty good for a ship—and amazing for one with her run of luck. The end came in May 1889. A salvage firm went to work on her with chisels, heavy mauls and a massive iron ball dropped from a height to jar her rivets loose. Even here she gave everyone a bit of a hard time. It took more than a year and a half to demolish her.

By way of a eulogy, let it be remembered that in her hard luck she made one generous gesture for the benefit of other ships: She discovered the rock perpetuating her name off eastern Long Island.

A Grisly Secret

It was discovered when dismantling of the *Great Eastern* entered its final stages. At one time, many years before, she whispered hints that she might divulge this terrible secret, but they went unheeded. So she carried it in her metal belly for more than thirty years.

Sometime just prior to one of her earliest sailings, rumors began circulating that a "basher" or riveter had been sealed alive inside her double hull. How these rumors started isn't clear. Maybe it was a missing-persons report, or perhaps word that a worker had heard muted pounding and distant cries. In any event, the rumors became a kind of ghost story that affixed itself to the ship—as if she needed it! Old-time mariners added this to their lengthening indictment of the *Great Eastern*. The basher's wraith was sure to haunt the liner, they said.

When only the *Great Eastern*'s bottom section remained, the rumors and ghost stories became the ghastly truth. In a space between the inner and outer hulls reposed the skeleton of a man. Near it lay the mortal remains of another individual, probably a riveter's helper or apprentice. In darkness they had pounded and shouted to make their presence known. Their calls went unanswered, and finally they died, entombed within the metal walls of the ship they helped to build.

11

Twin Mysteries

Some sea riddles receive a lot of attention, whereas others always remain in the background. Yet these unheralded mysteries are just as fascinating. Such obscurity cloaks the case of the *Beach Bird*.* The "evaporation" of her captain and crew was every bit as mysterious as the disappearance of those aboard the *Mary Celeste* and *Carroll Deering*, but has been seldom mentioned.

The brig *Beach Bird* was en route home to Rhode Island from a trading voyage to Honduras, laden with an exchange cargo. So far as is known, it was an uneventful return run—at least up to a point. Yet she was overdue. When her delay lengthened, her owners' anxiety turned to fear; but this was mollified temporarily by a report from another vessel that the *Beach Bird* had been seen down the coast, perhaps two days away from Newport. Her owners couldn't fathom the delay, but they took up a vantage point on shore and kept a vigil, nervously scanning the watery distance with telescopes to watch for the familiar hull and rigging of their ship.

The report that the *Beach Bird* was only two days from Newport turned out to be erroneous. Very erroneous. Days passed without either sight or word of the missing brig. Hope faded. Just as her owners had about resigned themselves to writing her off as a total loss, they were informed that she had been seen approaching Bentons Point, a landfall near Newport.

By now practically the entire town had heard about the long-overdue ship. People hurried to the harbor to hear what they were sure would be an exciting account of her delay. Imagine their surprise, and her owners' consternation, when the *Beach Bird* sailed blithely by Newport harbor entrance and headed straight for a rocky area. While villagers looked on in bewilderment and her owners pulled their hair out by the roots, the brig deftly altered course, barely enough to escape the rocks, then went hard aground on a sandy beach.

Spectators rushed to the scene of the beaching, and some climbed

---

*She may be the vessel mentioned as *Seabird* in some accounts. There's confusion in that detail.

aboard the stranded brig. She was searched minutely from stem to stern—deck, cabins, hold, even lockers and dark corners. Her cargo seemed to be intact. A fire blazed in the galley stove, a steaming kettle had been atop it, and a table was set for a meal. But not one human being, alive or dead, was on board. The only signs of life were the vessel's mascots, a dog and cat, quite well but wondering what was going on. Unfortunately, they couldn't answer questions.

No survivors or bodies from the *Beach Bird* were ever found. Nor did any clue to the cause of the disappearance of her men ever come to light. It was as though some supernatural force guided her home.

Apparently her cargo was salvageable, but she was not. *Beach Bird* was left to the slow ravages of time and weather. Then, during a storm, her remains disappeared. (One might ask, was it to rejoin her crew?)

The Other "Twin"

Beach Bird had a Pacific Coast counterpart in the schooner *J. C. Cousins*.

In the 1880s the handsome *J. C. Cousins* served as a pilot vessel, whose area was the mouth of the Columbia River in Oregon. Her responsibility was to safely guide incoming ships past dangerous sand bars and shoals in the river's mouth and estuary. Her master was Captain Alonzo Zeiber of Astoria, Oregon.

On a crisp but fair October morning in 1883 the trim 87-footer received orders to take up a station outside the river mouth and await an incoming French ship due to arrive the next day, then guide her into the harbor. Since the *J. C. Cousins* drew a bit of water herself, and the tide was not yet right, she dropped her hook in the estuary to wait for sufficient depth on the sand bars that lay between her and the open ocean. In light of subsequent events, this pause was important because the schooner could be observed by other vessels and people on shore. Nothing was amiss.

Later that day the tide was right. The *J. C. Cousins* glided over the treacherous shoals and took up station on the open sea beyond the Columbia's mouth. She was seen there that night by men in a nearby lighthouse. Again, all appeared to be well aboard the vessel.

At dawn the next day, with weather still crystal-clear, the pilot schooner was seen tacking to seaward, presumably for a rendezvous with her French client. (I say "presumably" because there's no mention of the other vessel having appeared.) Then, abruptly, while

86

witnesses watched in open-mouthed disbelief, the *J. C. Cousins* altered course from offshore to inshore and headed directly for a sprawling sand bar. As a pilot, Captain Zeiber knew about this bar, of course, and even a stranger would have been warned by the lighter-colored water characterizing shoals and by the waves breaking on them. Yet the schooner steered for certain calamity without hesitation.

Dumbfounded observers watched her crash into the bar with great force, the impact almost enough to dislodge her masts. And there she became wedged, incoming rollers breaking over her stern. Now telescopes focused on the stricken schooner. They saw no one on her decks, no frantic efforts to launch lifeboats, no one jumping into the water or already there. Nor had the vessel been in trouble earlier, so far as had been seen. She flew no distress signals.

Surfboats were launched by men of the U.S. Lifesaving Service (later incorporated in the Coast Guard). Their calls to the schooner went unanswered as they approached. The reason became obvious when they got aboard. Captain Zeiber and three crewmen known to have been on that assignment were gone, with no clues as to why, how or when. The schooner was still sound, despite her violent collision with the sand bar. Her rudder and steering apparatus were functional, which deepened the mystery of why she suddenly put about and headed for grief. Her lifeboats were still in their davits.

Also adding to the enigma, everything was in order aboard; and in his last scrawled entry in the log, written that morning, Captain Zeiber noted that all was well. Obviously, it didn't remain that way. And it would seem that whatever happened to the men on the *J. C. Cousins* must have occurred not too long before her suicide, because there was still a fire in the galley stove, and untouched food remained on a cabin table. Yet, in another contradiction, there was no disarray in the crew's quarters, such as might have occurred if they had had to leave in haste. For that matter, there were no signs of disturbance anywhere on board. Piracy had to be ruled out anyway; and a mutiny was extremely unlikely because all hands were known to get along together at work.

In the hours and days following the schooner's grounding, searches failed to turn up any survivors, corpses washed up on beaches, or positive clues to what happened aboard the *J. C. Cousins*. As in the case of the *Beach Bird*, it remains an unsolved mystery.

Also as in the case of the Rhode Island brig, the schooner could not be saved. She carried no cargo, of course, but some of her equipment, hardware, and other items were salvaged. Then she too was erased for all time by waves, weather, and winds.

The *J. C. Cousins* left an eerie postscript, however. For a long time afterward rumors persisted that Captain Alonzo Zeiber had been spotted in various distant ports of the world. They turned out to be only romantic rumors. The one undeniable fact was that the schooner's men had vanished forever, removed by cause or causes unknown.

12

"Who Are We Having for Dinner?":
Cannibalism at Sea

Throughout history, relatively few civilized people have eaten human flesh. But sea history is spiced by a number of recorded instances in which the gaunt specter of starvation forced mariners into considering shipmates as the main course.

A Whale Did 'em In

What could be termed a classic case of cannibalism at sea is that involving the whaler *Essex* in the South Pacific back in 1819.

With discovery of petroleum yet to come, whale oil was in large demand during the first half of the 1800s as an illuminant for lamps and as an excellent lubricant for fine machinery.* U.S. whaling ships sailed from ports such as New Bedford, Massachusetts, and Sag Harbor, New York, to hunt the giant marine mammals all over the globe for their oil, often on voyages that kept them away for as long as three or four years.

Whaling ships were literally floating "refineries." After hand-harpooning by the mother vessel's whaleboats, the carcasses were brought alongside and flensed, or stripped of their blubber. The blubber was cut into chunks and "tried"—cooked to extra its oil—in huge kettles right on the whaler's deck. After the oil was rendered from the blubber it was stored in barrels in the ship's hold until return to port.

The most highly prized targets were sperm whales or cachalots. Not only does this animal's blubber have a high yield of oil, but carries in its vaulted head a reservoir of oil of extremely high quality.** Called spermaceti to differentiate it from blubber oil,

---

*Until its ban in the United States as a conservation measure, whale oil bridged the past and present by being used in aircraft engines.

**A theory now is that this reservoir of oil functions in some way to enable a whale to dive to great depths and return to the surface quickly, all without suffering the "bends" or other injury.

this liquid brought a higher price as a lubricant for clocks, watches and other delicate machinery.

The financial rewards could be greater than those from other species, but so could the risks when tangling with sperm whales. Notoriously short-tempered, uncooperative and particularly resentful of being stabbed with harpoons, cachalots smashed many a whaleboat to splinters and cost numerous hunters their lives. Moreover, they also were known to charge whaling ships when sufficiently aroused, sometimes with disastrous results.

Wild Justice

Such was the destiny of the whaler *Essex* on the South Pacific's broad reaches in 1819. The vessel was charged by an especially large and very irate cachalot. So great was the impact that the whaler's timbers were stove in, and she sank quickly. The *Essex* went down so fast, in fact, that her crew barely had time to launch her smallboats and get clear.

They survived the sinking, but the reprieve from death looked like a short one. They were adrift on the largest and loneliest of oceans, with nothing more than a prayer for hope of rescue. And their hasty departure from the *Essex* had allowed no time to bring provisions and water.

As days and nights crawled by, wind, currents, and darkness separated the smallboats, now left to face death in whatever form it might appear. Starvation was one; without rain, thirst was another.

Some of the survivors held starvation at arm's length by resorting to cannibalism. It isn't known if they waited until their fellow crew members died from natural causes or drew lots to see who would be sacrificed; but the flesh and organs of the departed sustained them, and blood may have helped to fend off thirst. In time, a smallboat carrying three *Essex* survivors, now little more than babbling subhumans, was rescued by the English brig *India*. Several weeks later, the whaler *Dauphin* came upon another smallboat carrying Captain Pollard, master of the *Essex*, and one seaman. Barely alive, they looked like bearded skeletons.

Horror Drama, Sea Style

One of the grimmest, ghastliest of all sea chronicles is a hideous drama that followed on the heels of the Napoleonic Wars. It's the story of the French frigate *Méduse*—or, in English, *Medusa*. In their contemporary version two survivors, Jean Baptiste Savigny and Alexandre Correard, wrote: ''Maritime annals have no greater example of a shipwreck than that of the frigate *Medusa*.''

In July of 1816 the 44-gun *Méduse*, a Captain La Chaumareys commanding, set sail for Africa to repossess the French colony of Senegal from the British. Aboard was a motley human menagerie. Swaggering about the ship were members of a military expedition: Tough, surly, battle-scarred veterans of Napoleon's army, giant black colonials and their officers. In obvious anticipation of victory, the passenger list also included a new governor for Senegal, a pompous, overfed individual with the improbable name of Schmaltz. Filed away in various accommodations were male and female settlers bound for a new life in the African colony, men from different trades and professions, students and drifters seeking adventure, assorted rogues, and even a few pimps and some *filles de joie* to establish a whorehouse or two in Senegal.

Bad Start, Then Trouble

For a while the voyage was only a comedy of errors. It almost came to an unscheduled early end when the *Méduse* ran aground in the Bay of Biscay. On the next leg, to the island of Madeira, she wandered a hundred miles off course. These errors become understandable when you learn that navigation of the frigate had been entrusted entirely to a passenger who knew little about the sea so that Captain La Chaumareys could luxuriate in his quarters with a lady and plenty of wine. It was merely a question of time before they all ran into real trouble. It came on July 2nd when the *Méduse* ran aground on Arguin Bank, sixty miles or so off the African coast. These shoals were well marked on charts, but the amateur navigator-skipper was playing it by ear.

Although the frigate was not damaged severely and there was no immediate peril, the grounding triggered instant panic and wild disorder among the passengers. The idiot who got them into this mess wisely abandoned the helm and hid; and the real captain, rudely yanked from his paramour's arms and somewhat the worse for wine, was of no help. He and incumbent governor Schmaltz refused to allow jettisoning of cannons and cargo that would have lightened the *Méduse* enough to float her free. Thanks to further incompetence by her skipper, who was unfit even when sober, the frigate nestled deeper into the shoals' sand. When seas made up the next day, she was doomed. When she started to come apart from the battering, it was decided to abandon ship.

Immediately there arose a problem in mathematics and physics: How to shoehorn 400-plus people into six lifeboats with a maximum capacity of 250? The answer is simple: It was impossible. Actually there were only four lifeboats for the common horde. Captain La Chaumareys took one of the best boats for himself and his mistress, as well as a supply of wine, provisions, and just enough of the crew for oarsmen. His boat was less than half-filled. Governor Schmaltz departed in similar luxury—in an armchair—with his family, their luggage, and ample provisions and liquid refreshments. His private barge also was less than half-filled. The other two boats were tossed to those remaining and after considerable tussling were quickly filled by the strongest passengers. Everybody else was left to his or her own fate.

Somehow, those left behind evolved a plan from chaos. They put together an emergency raft from the stricken *Méduse*'s masts and spars, with planks from her splitting decks as a crude floor. Measuring roughly sixty-five feet long by twenty-three wide, its capacity was calculated—or, rather, miscalculated—at 200 persons, plus whatever provisions they could take along. The boats would tow this contraption to shore.

Fair weather and a calm sea blessed the day of departure; it was to be the only good break they would get. Some 147 people crowded aboard the raft. Among them were soldiers and their officers, passengers, sailors, and a lone woman (presumably the *filles de joie* were thoughtfully taken in the boats). With them they had several barrels of flour, two small casks of water and six barrels of wine. So crowded was the raft that its occupants had to stand shoulder to shoulder. Under their weight it submerged until waves lapped at their thighs. Many of the men were still drunk from a binge on the *Méduse* the night before.

As agreed, the boats began towing this overloaded vehicle toward the African coast, far out of sight over the horizon. (Under the circumstances, it's a wonder anyone knew which direction to go.) But when they discovered what a monumental, back-wrenching chore this was, the towing boats had a change of heart and released the raft. One hundred and forty-seven people were left standing knee-deep in water on a precarious contrivance that had no means of propulsion or control, abandoned to the caprices of weather, wind, and currents. Having had to jettison their barrels of flour to lighten the raft, the unfortunates were left with only about twenty-five pounds of biscuits, already soggy, and their pathetically inadequate water supply—and six barrels of wine. The nearest coast was little more than fifty miles distant, but it might as well have been five hundred. And they were at the point of no return to the *Méduse*. They could still see her, but wind and current were widening the gap.

Their first meal was their last. Twenty-five pounds of biscuits, however rationed, do not go very far among 147 people. Wine provided a little sustenance.

Then came nightfall. Perversely, waves grew in height and the wind freshened. The ungainly raft rolled and lurched underfoot. As its improvised lashings stretched, it changed shape, alternately creating openings into the black water, then closing them like a vise. Passengers were thrown about wildly. While water swirled around their waists, they clung to the raft's makeshift mast and each other to keep from being washed or thrown overboard. Several didn't make it. In that nightmarish situation it was impossible to tell where the raft ended, until it was too late. Others fell into openings when the raft expanded and were trapped to be crushed or drowned when it contracted.

By morning of the second day there were 126 left. Mercifully, they didn't know yet that the horrors would intensify.

With darkness came greater terrors. Again the wind freshened and waves heightened, sweeping across the raft and tossing spume in their faces. Every minute their raft's antics threatened to toss others into the sea—and did. Fearful of disturbing the platform's overloaded balance, yet unsure of their footing, as many as could jammed themselves together in the center section so tightly that some thought they would suffocate. Those who couldn't be accommodated there were washed overboard. Many died in the sea that night. In a litany of shrieks, moans from the injured, and a howling wind, the survivors clung to anything, and never expected to see another dawn.

To compound the horror, soldiers and seamen broke into the wine barrels and drank themselves into a mindless state. One man went stark raving mad and began hacking at the raft's ropes with an axe. He was promptly executed as a matter of self-preservation. Then the soldiers, half out of their minds from terror and drink, unleashed pent-up hatred of their officers. The revolt became a battle royal. Back and forth on the rocking raft the bloody brawl raged. Slashing sabers caught fighters and noncombatants indiscriminately, killing several and injuring others. Suddenly yawning holes underfoot claimed additional victims. The raft's lone woman and her husband were tossed overboard by the mutineers, but were rescued. Weather cleared and waves abated as the moon shone on a temporary truce. Before long, the soldiers were attacking their officers again.

Sixty died that night, most of them soldiers. Numerous survivors suffered wounds or injuries of various kinds.

A hot African sun rose on a calm sea at dawn of the third day.

Most of the survivors were beyond caring. Some bled into the water from ugly saber wounds acquired during the previous night's free-for-all. Others had broken bones, severe contusions and bleeding scrapes, casualties from falls into the raft's interspaces. Sharks circled them relentlessly, drawn by the scent of blood and a trail of bodies.

Most insidious of all, two unseen presences had joined the raft's company: Thirst and starvation. With their meager water and food supply long since consumed, only one barrel of wine remained. A few men tried to catch fish with crudely improvised gear, but failed. The remaining soldiers gnawed at the leather of their belts.

Its passenger list sharply reduced, the raft now rode higher in the water. Its scene combined a madhouse and a human abattoir. Corpses from the previous day's fray lay about. Sight of them inspired a hideous act for survival.

When he could no longer stand the pangs of hunger, a soldier began cutting flesh from one of the bodies. Similarly maddened by the nagging of their empty stomachs, several joined him in the horrid feast. Other looked on incredulously, torn between revulsion and their own starvation. At first they couldn't bring themselves to partake, but in time the all-powerful urge to survive overcame their disgust. Thirst went unslaked by rationed mouthfuls of wine.

On the night of the third day twelve more died.

July 8th was their fourth day of torment. Now forty-eight remained, clinging to life and sanity only by slender threads. A fragment of good fortune came this day when some fish swam under the raft and were trapped in its spaces. But it was a poor windfall. The fish were small and the survivors were famished. Again they turned to strips of human flesh, this time from a corpse held from the previous night.

For a while things had been peaceful on the raft, relatively speaking. But that night, for reasons known only to themselves, the soldiers again set upon their officers. Skeletonlike figures grappled and slashed at each other in the moonlight. By dawn eighteen more men were dead, and all the other combatants had additional wounds. The poor woman, who had had rough going all around, again was thrown into the sea, but was rescued. By then she must have been wondering which fate was worse.

Still No Mercy

On July 9th, fifth day of their suffering, the big raft continued drifting, its occupants neither knowing nor caring where. So far as they were concerned, they were in hell already. New tortures were added to old as wounds festered and sores developed on their legs, keeping them in continuous pain. "Only twenty of us were able to stand upright or move about," wrote Savigny and Correard in their account a year later, ". . . we had wine for only four days."

By the sixth day they were subsisting solely on wine, now even more severely rationed from what was left in the single remaining barrel. Two soldiers who attempted to take more than their share were summarily executed by being shoved overboard to waiting sharks. During the day a young lad died.

Twenty-seven remained, miraculously holding onto life in thirst, hunger, pain, and the African sun's merciless burning. But among them a dozen were dying. One was the courageous woman who survived two forceful dunkings in addition to her other woes. Her legs were swollen far beyond normal girth, their skin shredded to her thighs. Eleven were little more than skin and bones, covered with suppurating sores and wounds, and slipping across the border between partial sanity and full lunacy.

The other fifteen held an emergency council. The supply of wine was dangerously low. Their twelve sick companions obviously were doomed. To keep them on such tiny rations of wine would serve only to prolong their misery and stave off death for just a few days at most. On the other hand, increased rations would not prevent the inevitable, but would only jeopardize whatever chances the others had. The council reached an agonizing decision: The twelve would be consigned to the sea, ending their suffering. Three seamen and a soldier, the only ones with enough nerve and strength, carried out the verdict. This time the unfortunate woman did not return. Waiting sharks were swift executioners. The council estimated that now they had enough wine to keep them alive for a few days longer.

On July 14th, back in France, it was Bastille Day, a national occasion for celebration after the French Revolution. Out on a now-roomy raft there was no holiday for fifteen inhuman-looking wretches, baked into a torpor by a merciless sun. Their stomachs were contracted in emptiness, but their parched tongues swelled and all but choked them. They were beyond feeling, beyond caring; yet a tiny flickering flame, a pilot light of life, refused to go out. By day they managed to find a little relief from the sun's heat by crawling into the shallow water covering the raft's forward section. By night under the stars they took turns resting on a makeshift platform out of reach of waves. They were so weakened that even the tiny sips of wine produced intoxication, and periodically someone had to be restrained from hurling himself to the patrolling sharks.

Incredibly, still more of this unspeakable torture was to follow before salvation. On July 17th a brig named *Argus* found them. The vessel had been sent from St. Louis, Senegal, on a search. As the brig drew alongside, pathetic figures crawled to the raft's edge. The faces that looked up were like skulls, eyes deep-set in dark hollows. Whimpering, they were lifted from their nightmare world, back into a more humane reality.

Epilogue

But the survivors would never be the same. The horrors of their thirteen days on the raft left them with irreparably damaged bodies and shattered minds. Five died soon after being brought into St. Louis. Only the grave ended the memories that haunted the others.

Not surprisingly, but in a miscarriage of justice, Governor Schmaltz and Captain La Chaumareys reached shore safely. Four boats in all made it.

Even Into the Twentieth Century

In the summer of 1904 a small fishing vessel sank during a storm off the coast of Maine. Among ten men aboard, only three were able to save themselves. Jeb Cannon, Jim Thomas, and Clem Mallory managed to ride out the blow in a lifeboat. When the fishing vessel failed to return to port, a wide search was launched, but no trace of her or her crew was found. What happened came out later when one of the survivors finally was picked up. His was a gruesome story.

About a month after the sinking, Jeb Cannon was found adrift in his lifeboat by a trawler. The physical change in the man was dramatic. His hair, once a youthful brown, had turned completely gray, contrasting with a face so deeply tanned by the sun that it almost looked charred. In the lifeboat with him were some unidentifiable chunks of meat.

Jeb became highly emotional when brought aboard the rescuing vessel, and it was only after some rest and comfort from gulps of rum that he calmed down sufficiently to tell his story. He had survived death from thirst by catching rain water in his shirt, he said, but it was with hesitancy that he told what he had done for food. Exposure eventually killed his unfortunate companions, he related: First Mallory, then Thomas. The unidentifiable chunks of meat in the lifeboat were Jim Thomas.

That story was not quite accurate, it turned out. The true version came as a deathbed confession fourteen years later.

When the three men hastily took to the lifeboat, Jeb said in a voice barely above a whisper, there was no food or water. For days they drifted aimlessly, suffering thirst and facing starvation. After nearly two weeks of this torture, Jeb shot Clem Mallory with a pistol he had taken with him. Why he chose Mallory over Jim Thomas, he didn't explain. Perhaps the candidate already was far gone. Jeb ordered his surviving companion to butcher the body so they would have food; but Jim refused, either because he had no stomach for the hideous assignment or because he wasn't in much better shape than the deceased. Cannon did the job himself. Until the meat spoiled and smelled so awful they had to get rid of it, there was something to eat.

Thus they were sustained for several days. Understandably, Jeb maintained a tight hold on his pistol whenever he slept. Any doubts that Jim may have had about his own future were cleared up abruptly three days before rescue by the fishing vessel. Jeb shot him for a new supply of meat.

Finale

The story has it that Jeb Cannon's fellow townspeople were so appalled by his deathbed confession that they refused to tender him the customary funeral and burial. Instead, they took his body out to sea and unceremoniously threw it overboard.

Beach Party

Autumn, 1710, saw the English merchantman *Nottingham Galley* set sail from Ireland with a cargo tagged for New England, to be unloaded in Boston. In charge was a Captain John Dean. Under him worked a crew of fourteen.

The Atlantic crossing was negotiated without noteworthy difficulty, but as the ship neared the Massachusetts seaboard she ran into a December nor'easter. Storm clouds having prevented her navigator from taking sightings, and wind-driven rain and snow drastically impairing visibility, the ship wasn't quite sure where she was. Notification came suddenly and rudely as she drove hard upon rocks at a deserted place called Boon Island, offshore of a coastal community now known as York Beach, Maine.

The *Nottingham Galley* quickly incurred fatal injuries. Her masts tumbled; and with her back broken by the rocks, she began to come apart under the waves' hammering. Captain Dean decided to abandon ship immediately. He and his men threw themselves into the sea, striking out as best they could for the island's shore. It's in the category of miracles that they all made the storm-battered beach safely.

But their ordeal was only beginning. They spent a miserable night huddling in the lee of rocks that afforded inadequate shelter against frigid wind and a mixture of snow and rain. By sunup the nor'easter had roared elsewhere, but they already were chilled to the marrow and it was bitter cold. Efforts to ignite a fire with flint and steel failed, any wood they could collect being wet. Shivering from cold, they gained a little warmth by actively searching for food and other items washed ashore from their late ship. Part of the *Nottingham Galley*'s cargo was cheese, taken aboard in Ireland. To their acute disappointment, about all of it that came ashore were some small chunks. There was nothing from the vessel's provisions.

Faith, Hope, and Frustrations

Perceiving that they were marooned on an island within sight of the mainland—about twelve miles, Captain Dean estimated—they took heart from the thought that they soon would be seen and rescued by a passing vessel. Meanwhile, they fashioned a crude tent from pieces of sails that had washed ashore. This at least gave some protection from the wind, but not from the cold, which intensified, threatening to cost them feet and hands from frostbite.

Scouting the beach was rewarded by finding timbers, a chest containing a few tools, and some caulking oakum, all from the *Nottingham Galley*. It was as much in an attempt to keep warm and divert their minds from their predicament as for a means of escape that the little band started to build a boat, a project hampered by numbing weather. Along about that time they experienced their first loss. The ship's cook died. His body was placed on the beach where tides could carry it away for disposition at sea. They had no means of digging a grave in the frozen ground. Later Captain Dean was to say that no one mentioned eating the late cook, but added that he and others had flirted with the idea.

Days passed without rescue. Once they saw three boats pass the rocky isle, but at that distance their frantic waving and yells went unnoticed. Now they had been marooned a week, barely subsisting on the salvaged chunks of cheese as long as they lasted, then on some beef bones stranded by a tide.

They managed to finish the boat with which they hoped to escape this forlorn place. Launching was accomplished only with great difficulty, every man being very weak, then was crowned by bitter disappointment. In the surf the boat capsized and was stove in by rocks. Their hopes gone and spirit broken, they were reduced to a most distressing existence. They were starving; feet and hands were frozen—some gangrenous; deep, running ulcers of the legs added to their agony; and they still had no fire for warmth. So far they had kept barely ahead of complete starvation by eating whatever sea-

98

weed and mussels they could gather among the rocks, tides permitting; but now, in their condition, even these few items were becoming very difficult to obtain. Once they struck down a seagull. The
bird provided only a tiny morsel for each man, but they were
grateful.

Another sailboat passed their island prison without seeing them.
In desperation the men mustered what little strength they had left to
fashion a crude raft on which they hoped to reach the mainland, so
tantalizingly in sight. Again they met failure in launching, and this
time it cost two lives. Days wore on with terrible slowness; each
night seemed endless. Attempts to kill sea birds with sticks and
stones failed. Lower tides were required to get at their dwindling
supply of mussels and seaweed.

Then, Finally . . .

One day the ship's carpenter died. He was still a big, heavy man,
and the survivors wondered how they would be able to transport his
body to the tide line, where waves could carry it away. When
Captain Dean ordered its removal from the tent they could barely
drag it outside. Now a dreadful question came out in the open: The
men asked Captain Dean if they could eat the corpse to stay alive.
Their commander wrestled long and hard with his conscience, and
in desperation gave his consent.

He ordered the decedent's head, hands, gangrenous feet, skin
and intestines to be removed and discarded in the sea. That accomplished, he directed that the body be quartered so that it would be
easier to transport and use. This gruesome chore fell to the captain,
the others professing weakness or protesting that they simply
couldn't do it. It was a difficult job in all ways, but by sundown it
was done. For their first cannibalistic meal they dined on slabs of
flesh, with seaweed as a vegetable. Three of the men chose impending starvation over this repast, but by the next morning they
could resist no longer.

In his account of their terrible sojourn on Boon Island, Captain
Dean noted that all hands, himself included, soon ate the carpenter's remains with such greed that he was obliged to remove them
from the others' reach, lest they quickly deplete the supply and
make themselves ill in the bargain. From then on Captain Dean had
to watch over the meat supply as though it were gold. And guarding
it became tougher as the men's personalities changed radically.
From peaceable, cheerful, obedient individuals they turned into
wild-eyed, rebellious barbarians, given to sudden, vicious quarrels.
No doubt only their physical weakness saved Captain Dean from

becoming the next supply of food. Just a few more days in that situation, he said, and there would have been no waiting for a corpse—they would have made one.

Luckily, they were rescued before that could happen. Ultimately they returned to England, where Captain Dean was charged by his first mate, one Christopher Langman, with severely beating members of the crew (early in that unfortunate voyage) and planning to deliberately wreck the *Nottingham Galley* for insurance.

Apparently these grave charges were never proved—or disproved.

13

Big Boom in Bombay

Paradoxically, one of maritime history's greater disasters is also one of the least known catastrophes of World War II. It was the explosion of the 7,142-ton cargo ship *Fort Stikine* in Bombay, India, on April 14, 1944.

The *Fort Stikine* had chugged away from England two months earlier with a mixed load that included explosives and ammunition destined for Allied supply depots in Bombay and the Asiatic war theater. After a stop at Karachi to discharge part of her cargo and replace it with a new lading that listed oil, resins, and upwards of 8,000 bales of cotton—all inflammable materials, it should be noted—she eased into Bombay harbor. And "eased" is the word. World War II was at white heat, with plans in the works that boded no good for Japan. As a gateway to Asia, Bombay harbor was literally wall-to-wall with Allied vessels of all kinds. The *Fort Stikine* elbowed her way in alongside a wharf between two other freighters, *Belray* and *Japalanda*.

First Warning

At noon on the fatal day dock workers broke for chow. Soon thereafter a crewman relaxing at the *Belray*'s rail noticed suspicious-looking smoke pouring from a ventilator on the *Fort Stikine*. For reasons unknown, no alarm was turned in. No alarm was sounded until stevedores returned from lunch and discovered a fire gnawing the contents of the *Fort Stikine*'s No. 2 hold. What followed was a series of errors in judgment and tragic inadequacies leading to cataclysmic disaster.

To begin with, the *Fort Stikine* was not flying the red flags required to be displayed on vessels carrying explosives. Had these been flown, neighboring ships would have moved to safety at the first signs of fire. Instead, they remained at their berths, unsuspecting, to watch the fire be put out. Meanwhile, there was difficulty with dockside communications in summoning fire-fighting apparatus. When it did finally arrive, nearly two hours after the *Belray* seaman first noticed the smoke, it was woefully inadequate. Additional men and equipment were summoned.

Soon there came another, more ominous warning, which no one interpreted and heeded. The color of the smoke now billowing from the *Fort Stikine* in increasing quantity and density had changed to an ugly, yellowish brown, signifying that the fire had reached the ship's cargo of explosives. She had, in effect, become a 400-foot-long bomb which could detonate at any second.

The fire had also reached the vessel's cargo of fish, taken on at Karachi; the ship's master, Captain A. J. Naismith, immediately concentrated on having it unloaded because of the terrible stench. Along about this time, an ordnance officer came aboard and advised Captain Naismith to scuttle the ship as soon as possible. In a counterproposal the dockmaster pointed out that Bombay harbor was too shallow for scuttling and recommended that the *Fort Stikine* be hurried out to sea. These contradicting suggestions only confused the skipper, and he went ashore to contact the freighter's insurance agent for advice. While all this was going on, fire crept closer to the explosives.

Kapow!

Without further warning, the *Fort Stikine* became a monstrous fireworks display. Skyrockets of sparks and fireball Roman candles shot into the sky, followed in seconds by an explosion that rocked the harbor and rattled windows miles away. This initial blast blew sixty-six firemen into eternity and seriously injured nearly 100 others. Minutes later came an even greater explosion. It tore the *Fort Stikine* apart, reducing her superstructure and hull to hot shrapnel that flew half a mile up and in all directions. Jagged pieces of metal and other lethal debris rained from the sky in a mile-wide circle.

Such was the second explosion's force that it generated a 50- to 60-foot wave which lifted the 5,000-ton *Japalanda* and dropped her on wharfside buildings. Docks and adjacent buildings not already leveled by the blast or flattened by the *Japalanda* were swept by fire. Within the cataclysm's wide range, numerous other structures, including business establishments and dwellings, likewise were destroyed. Those beyond the outer perimeter had their windows blown out or broken. Entire wharves had vanished.

The Costs

In a country as populous as India and in a city as large and busy as Bombay, it was impossible to determine the number of human victims with any accuracy. The toll could only be approximated by a count of those still around to be treated at hospitals or sent to morgues. An admittedly incomplete count of fatalities stood at 1,500, to which was added a list of 3,000 or more injured. Since there could be no complete census of people living, loitering, and working in the disaster area, the full death toll will never be known. Quite possibly another 1,000 or more were blown to pieces by the blasts or cremated in the fierce shoreside fires that followed.

All items on the bill for the *Fort Stikine* disaster were stunning. An uncountable number of docks, buildings, and other structures had either been erased or destroyed. More than twenty-five ships were sunk by the explosion's "tidal wave" or gutted by fire. Loss of Allied shipping stood at about 100,000 tons, with a price tag well above one *billion* dollars. The loss of small boats, barges, and the like couldn't even be estimated.

Added were the costs, in money and precious war time, of cleaning up the mess. Even with a work force of ten thousand men and their equipment, some six months were needed to rid Bombay harbor of the shattered remains of ships and other debris, estimated to have totaled upwards of one million tons.

The cause of the fire that started it all, in the *Fort Stikine*'s No. 2 hold, was never determined. A logical suspicion is sabotage, which it undoubtedly was, but proof positive was never found—or at least never announced publicly.

14

Sea Superstitions

Mariners' superstitions should be included among mysteries of the sea if for no other reason that the fact that their origins are largely unknown. It's also a mystery to us now why many of them were believed. Besides, they form colorful threads in the tapestry of maritime history.

You can be sure that these superstitions were believed wholeheartedly by old-time seafarers. In fact, I can't think of any other group of working men as superstitious as those sailors of long ago. The list of their odd notions—accepted as gospel truth, mind you—is as long as your arm, maybe longer. Some were practically universal, shared wherever seamen gathered in fo'c'sles. Others were peculiar to certain countries, or even regions.

By and large, old-time mariners were a pessimistic bunch; and we can't tell which came first, their superstitious nature or their pessimism. Most of the superstitions involved bad news of one kind or another. Relatively few of those I've run across say anything about good luck. I think I know why. The old-timers spent much of their life at sea amid potential dangers, if not outright peril, and faced death in many forms. That's hardly conducive to optimism. Superstitions were expressions of their fears. And maybe it was because their fears were based on reality that their superstitions were borne out at times.

Way, Way Back

If you remember your history in school you'll recall that for centuries, on into the Middle Ages and later, people in Europe were sure the world was as flat as a billiard table. What could you expect? In those dimly illuminated times they also thought that eclipses meant the end of the world and that animal life could spring spontaneously from inanimate matter such as dust and dew. People with ideas like that could believe anything.

If one were to sail far enough, the world-is-as-flat-as-a-table theory continued, one would fall right off the end. Nobody knew exactly what happened then, but it was enough to give mariners the galloping fantods every time they ventured any distance at sea.

104

Fortunately, there were some cooler heads around Europe in those days: Sensible thinkers, mathematicians and astronomers. They may have risked torture and execution for heresy, witchcraft or being dangerous cuckoos, but they broadcast a theory that the world was round. When they found subscribers in men like Christopher Columbus and other brassy navigators who went out and proved that theory, it was a big step forward.

That flat-world superstition probably slowed civilization's advance by at least a thousand years.

But proving the earth's roundness didn't clear up all problems. The Atlantic Ocean was still looked upon by European mariners as the Sea of Darkness, where only evil fates could, and probably would, befall anyone foolish enough to defy it. Moreover, mariners' charts as late as the 1600s graphically pictured seas and oceans teeming bumper to bumper with menacing monsters, the likes of which you wouldn't believe.

No one really had seen the nightmarish creatures pictured in sea books of the 1500s and 1600s, like the large monster called the "rosmarine," which had a horselike head, atop which were two short funnels spouting water. But people were great believers in mythology in those days, and they made up for what they lacked in knowledge by wild imaginations. Belief in sea monsters had been started centuries earlier by then-prominent men whose opinions were still respected. Herodotus, for example, a widely traveled Greek historian of the fifth century B.C., was considered a reliable reporter. People continued to believe his writings about such sea monsters as the one-eyed beast that was half fish and half lion, another that had a dog's head, another that had no head at all but wore a single eye in the middle of its chest, and one that was part eagle and part sea serpent.

No wonder later mariners developed fears and superstitions. Even Columbus, who was pretty level-headed, must have suffered terrors more than once as he crossed the Sea of Darkness to the New World.

Come to think of it, many of us still believe in such horrors today in the form of the so-called Bermuda Triangle.

Davy Jones' Locker

The sea has spawned a mythology all its own. An ancient, universal belief has it that the kingdom beneath the waves was ruled by gods, variously known as Neptune, Triton, Poseidon, and other names in various tongues throughout the world. To the domains of these sea gods were consigned the souls of mariners and seacoast dwellers who perished by drowning. Among American and British

105

seafarers the place where these souls were taken came to be known as "Davy Jones' Locker."

The true origin of this ancient expression is unknown. The *Oxford Universal Dictionary* traces it back to British nautical jargon of the mid-1700s. Other sources aren't sure of its origin, but suggest that it stems from "Deva Lokka," Hindu goddess of death. Or from a corruption of "Dyved," an obscure legendary Welshman. Or from "Duffy," the anglicized name of a West Indian spirit, while "Jones" was borrowed from "Jonah," whose "locker" was the belly of a whale. The last seems unfair to Jonah, since he eventually escaped.

In any event, the mythical Davy Jones came to be considered as a kind of sea devil, and his infamous locker as one of the sea's evil bailiwicks.

About Sharks

Men have feared sharks* for as long as they have braved the sea. It's understandable. Many a shipwrecked sailor saw companions yanked below the surface or mutilated by the toothy beasts. It was natural that their fears should give rise to superstitions.

Among some mariners the mere appearance of a shark was an evil omen, presaging disaster. If one followed a ship, it meant that somebody aboard was going to die. It didn't occur to believers that the shark may have been trailing the vessel because the cook or a galley mate was tossing garbage overboard. Nor did it help dispel superstition when sharks occasionally bit off the propellerlike device towed astern to measure a ship's speed.

Sharks have figured prominently in the lore and legends of island and coastal peoples throughout the world for centuries. Variously considered gods or demons, they were much feared either way, and gave rise to countless superstitions. Among some South Pacific islanders sharks were looked upon as deities that had to be appeased by human sacrifices. Adults and children were tossed to the beasts. After visiting the Hawaiian Islands in the 1800s, Russian explorer Otto von Katzebue described a large pen, about where Pearl Harbor is now, into which human sacrifices once were flung, to be torn apart by captive sharks.

Many religious beliefs developed around sharks in South Pacific islands. In ancient Hawaii there was a strong belief that the fish could assume human form—*mano kanaka*, "shark men," they

---

*The *Oxford Universal Dictionary* traces the term "shark" back to 1569, but its time and place of origin are lost in obscurity. Many years ago it was suggested that it stemmed from *schurke*, an old German word meaning "villain."

called them—and come ashore to commit murder or other mischief. Among other South Pacific islanders' beliefs the reverse occurred, with men taking the forms of sharks after death. Irish folklore has a tale about a man who was particularly evil in life and after death was reincarnated as a large, vicious-looking shark, condemned to cruise back and forth off the southern coast of Ireland for seven years as punishment. The legend said that this shark always appeared just before a severe storm.

Clergymen and Women

I could never fathom why, but it was a superstition among some mariners that to have a clergyman aboard was to invite a jinx that could spell potential disaster. In England it was considered a bad omen if the last person you saw before boarding a ship was a man of the cloth. Landlubbers would think the opposite, that to see a clergyman before boarding ship was a good omen and that to have a reverend aboard would be one of the stronger bids for protection.

As everyone knows, sailors have always been fond of women. But old-timers restricted that fondness to shoreside activities. When it came to having them aboard ship, women were viewed in the same unholy light as clergymen. It was bad luck if one attempted to board just before embarking, even worse if she booked passage. I don't know how old-time seafarers felt when ships carried women passengers; but they could cite instances in which vessels met disaster even with only one woman aboard—and sometimes she was the captain's wife at that.

Cats and Other Animals

In the old days aboard some vessels, notably American, cats were generally considered as unlucky additions, in spite of any good work they may have done in keeping down a vessel's mouse and rat population. If a cat washed behind its ears—as cats usually do—it was a sure sign of rain. If a cat or cats frolicked, a storm was sure to come. That they frolicked because they felt good was ignored. Worst of all, if an energetic cat mistook a mast or rigging for a tree and climbed it—look out!—the ship was headed for disaster.

I don't think I've ever heard any evil superstitions involving dogs. But that's understandable. People are more inclined to view cats with suspicion.

I've known professional skippers of sport fishing boats who do not look upon porpoises—also called dolphins—too kindly, claiming that they have seen them wantonly destroy fish. But in seafarers' lore porpoises and dolphins are signs of good luck.

A common belief, and not only among mariners, is that porpoises

107

keep sharks away, since these animals and fish are traditionally considered to be mortal enemies. In truth the belief can't be counted on. Commercial fishermen can tell you about seeing porpoises and sharks mingling peaceably on the same feeding grounds. And I've caught sharks in areas harboring both.

Propoises (or dolphins) are highly intelligent animals; and their friendly, happy-go-lucky deportment is probably why their sudden appearance around a ship was considered a harbinger of good fortune. Often these marine mammals will appear out of nowhere to sociably swim like a convoy for miles alongside a ship or sport fishing cruiser. Whether or not this is a good omen is a moot point (I've come back to the dock fishless on days when there were porpoises all over the place), but at least it's entertaining.

Old-time seafarers interpreted porpoises' antics in terms of weather forecasting. If they were playing in a certain way or milling about wildly, it foretold a storm in the offing. On the other hand, if they swam calmly through rough weather, clearing was sure to follow. Such interpretations might very well have some basis in fact. Marine creatures do react to weather changes.

Sea lore is studded with incidents in which dolphins purportedly guided or even "nudged" floating sailors toward shore. The authenticity of these stories is difficult or impossible to determine, but during World War II there was a supposedly documented case to support the belief. According to the story, some downed Allied fliers in an inflatable liferaft on the Pacific were repeatedly pushed toward the nearest shore by a porpoise. But it was a situation like that in the joke about Boy Scouts helping a little old lady across the street when she didn't want to go. The shore toward which the porpoise pushed the raft was an island occupied by the Japanese military. Maybe the porpoise was a Japanese ally.

Birds and Insects

Of all the different kinds of sea birds—gulls, terns, gannets, frigate birds—that figure in seafarers' superstitions, historically the most prominent is the albatross, immortalized among landlubbers by Samuel Taylor Coleridge's epic poem *The Ancient Mariner*. Somewhere, way back, there was born a strong superstition that to harm or, worse yet, kill an albatross brought a sure-fire ticket to perdition. The Ancient Mariner finds it out. "With my cross-bow I shot the Albatross," he confesses. And thereafter his vessel's luck is all the worst kind. She is becalmed, "As idle as a painted ship/Upon a painted ocean." The old sailor suffers terrible thirst under "a hot and copper sky," in which a "bloody Sun" blazes mercilessly. "Water, water, every . . . where,/And all the boards did shrink;/Water, water every . . . where./Nor any drop to drink,'' croaks The Ancient Mariner.

During the 1950s two real-life cases seemed to support the albatross superstition.

One involved a British freighter, *Calpean Star*. She was carrying a collection of birds and animals destined for a zoo in Germany, and in the menagerie was an albatross. A well-meaning but obviously uninformed crewman fed the albatross something that disagreed violently with the bird, so much so that it died. The curse didn't strike until later, when the *Calpean Star* was on assignment with a Norwegian whaling fleet. She was hit by a parade of troubles that included failure of her generators, a disabled engine, the ruin of the water supply by diesel fuel seepage, and damage to her rudder. Finally having to be towed into Montevideo, Uruguay, she ran aground, had an explosion in her engine room, and lost a crew member to drowning.

One might say, ''Well, maybe those things would have happened anyway, albatross or no albatross.'' Maybe so, but the freighter's crew wasn't sticking around for more. They flew home to England. The albatross curse seemingly wasn't through with them yet. They narrowly escaped disaster when their aircraft was damaged while landing in Rio de Janeiro.

Starring in the other tale of woe was the U.S. Fish and Wildlife Service research ship *John N. Cobb*. While she was off Cape Flattery, Washington, an albatross was sighted, and a member of her group of scientists requested permission to shoot the bird as a collection specimen. Permission was denied at first—possibly with *The Ancient Mariner* in mind—but finally granted. The albatross was shot down, after which the vessel suffered a series of mishaps that included destruction of an expensive net, breakage of the main winch's shaft, badly fouled net cables taking hours to retrieve, and injury to a researcher who fell aboard ship and fractured his ribs. And the fellow who shot the albatross became violently seasick for the first time in his life, even though the Pacific was glassy calm.

On the other hand, there's at least one entry on the good-luck side of the ledger of albatross legends. In 1956 a crewman on the liner *Southern Cross* fell overboard at sea and wasn't missed until his ship was some twenty miles distant. (That he was missed so soon was the beginning of his good fortune.) When the *Southern Cross* turned back to search for him, he probably would not have been found if the liner's lookouts hadn't spotted an especially large albatross circling and swooping over an object—the missing seaman—floating amid the waves.

Insects do not figure prominently in maritime superstitions, probably because they're land-based creatures and ships usually are beyond their range. However, it was held that a bee found aboard ship is a special sign of good luck.

Fleas, ticks or bedbugs carried aboard by sailors after shoreside excursions would be considered more in the nature of unwanted pets than superstition.

About Corpses and Others

Not surprisingly, the presence of a corpse aboard ship was viewed with general uneasiness, if not outright fear of what it might portend. Accordingly, any seaman or officer who died during a voyage was promptly sewn inside a weighted canvas sack and committed to the deep, with a few parting words said by the skipper. Of course, there was a practical public health angle to this too. On long trips without refrigeration a corpse could get pretty ripe, especially in the tropics.

A stiff was brought home only if absolutely necessary for some reason, in which case it must lie athwartship—that is, from side to side—and not in the vessel's long axis.

Bad luck on a voyage was almost certain to follow if a crewman saw a cross-eyed individual or a mentally defective person just before embarking. Ditto a lawyer or a tailor—or a clergyman.

Even the Bible

As strange as the superstition about men of the cloth being aboard was the belief in some quarters that misfortune would follow the quoting of Bible passages at sea. Except during burials, that is, when a suspension of belief was granted automatically.

Origin of the Bible-quoting fear is lost in antiquity; but, as often happened with seafarers' superstitions, it seemingly received enough support to keep it alive for generations. One of the more dramatic examples occurred far back in the days of sailing ships.

In this instance the destruction of three vessels off Britain's Scilly Isles in the early 1700s was blamed on a quoting of the 109th Psalm, which subsequently came to be dubbed ''The Cursing Psalm.'' The fact that the real cause may have been someone's incompetence was ignored by superstitious sailors.

The Scilly Isles were long infamous in shipping because of their dangerous rocks, on which more than a few vessels came to grief. According to a story widely circulated for years afterward, a seaman aboard the flagship of an admiral with the odd name of Shovel—Sir Cloudesley Shovel—had the temerity to warn Mr. Shovel about treacherous rocks in the Scilly Isles area they were cruising. Far from welcoming a helpful bit of intelligence, the admiral flew into a rage because a mere sailor had the brass to tell *him* how to navigate. He became so furious that he ordered the poor wretch to be hanged from a yardarm.

In his final moments before execution the hapless sailor bestowed a curse on Admiral Shovel and the fleet by reciting his own version of Psalm 109, which in essence said that the commanding officer should drop dead and be forgotten by all. Soon afterward, if we can believe the story, three vessels of the fleet, including the admiral's flagship, H.M.S. *Association*, met destruction on the Scillys' rocks.

Nazi Jonah

The superstition that a vessel can be hexed or jinxed didn't die with the era of sailing ships. It had strong supporters among workmen who constructed the Nazi Navy's 32,000-ton cruiser *Scharnhorst*. Her reputation for carrying a Jonah became so widespread in German naval circles that men were reluctant to serve aboard her (but we can be certain they had no choice).

The *Scharnhorst* was engineered as a superb, highly efficient fighting machine, fast and armed with extra-long-range guns. However, midway through her construction evil things began happening to give her builders pause for thought. At one stage she broke free from supports, squashing or otherwise eliminating some sixty laborers and injuring more than a hundred others. Shipyard men promptly declared her jinxed, and it became harder to draft work gangs.

After that calamity she was righted and construction continued until she was ready for launching. The Third Reich was anxious to exhibit the *Scharnhorst*'s might to the world. Specifically, the German Navy was eager to turn her loose on Allied opponents. What they didn't realize, or chose to ignore, was that her hoodoo was working for the enemy.

The *Scharnhorst*'s launching was to be an ostentatious affair, typical of Nazi pomp and circumstance. It would be attended by a delegation of the highest officials, led by no less than *Der Feuhrer* himself, with his henchmen Goering, Goebbels, and Himmler in tow. Only trouble was, the supercruiser couldn't wait. On the night before the scheduled ceremony she broke loose from her construction cradle and launched herself, destroying several barges in the process.

In Action

Once she got to fighting, the *Scharnhorst* lived up to her government's expectations. But a Jonah was still part of her crew. During a

111

devastating shelling of Danzig, one of her powerful guns exploded, killing nine sailors. In another gun turret the air supply system failed, suffocating twelve more men.

In the German campaign against Norway, during which she bombarded Oslo with her long-range armament, the *Scharnhorst* collected several enemy shells and was damaged seriously. A sister cruiser, the *Gneisenau*, towed her to safety and saved her from total loss. This near brush with eternity was a tribulation of war, but her hex reappeared not long afterward. Limping home for repairs, she hid by day and traveled at night under a protective mantle of darkness. On a particularly black night she crept into the mouth of the Elbe River en route to her dock. At that critical time her radar—her eyes in the dark—failed, and she collided with the liner *Bremen*, one of the world's largest passenger ships and the pride of the German merchant marine prior to World War II. The *Bremen* suffered more than the *Scharnhorst* in this violent rendezvous and settled into the Elbe's mud.

Back At It Again

For several months the *Scharnhorst* remained in port, temporarily removed as a threat to the Allies as she underwent extensive repairs and refitting. As soon as possible she left the sanctuary of the Elbe again and put to sea on missions of mayhem.

Forays took her north off the coast of Norway to prey upon Allied convoys. In darkness one night she unknowingly passed a British patrol boat, momentarily disabled by engine trouble. Fortunately, the patrol boat saw and identified the *Scharnhorst* and radioed an alarm. British warships hurried to intercept. They found the German cruiser and a lively duel ensued. For the moment, her Jonah was off duty. With her speed she easily left her attackers astern in the darkness.

But within minutes the *Scharnhorst*'s hex returned to active duty. Purely on a long shot one British warship let go with a salvo at 16,000 yards. The German cruiser's Jonah had placed her right smack in the line of fire, and she shuddered under thunderous hits. Flames leaped skyward, making the ill-fated cruiser a lighted target. Soon it was all over. The *Scharnhorst* was removed permanently from the war arena by a frigid sea. Most of her officers and men were eliminated with her. Fewer than forty survived.

The supercruiser's jinx then seemingly abandoned her to accompany two survivors. A postscript has it that they were killed by an explosion of their emergency heater.

Islands, 340 miles off the coast of Chile. There Selkirk quarreled bitterly with Captain Dampier and asked to be put ashore. His captain obliged, and he was marooned alone on little Mas-a-Tierra, an island fourteen miles long by eight miles wide. History records that he had a gun and ammunition, but little else.

For four years and four months Selkirk lived by his wits and ingenuity on Mas-a-Tierra in solitary exile. He had no native "man Friday" such as was created by Daniel Defoe so that Robinson Crusoe would have someone to talk to. In February 1709, he was taken off the little island by Captain Thomas Dover, master of a prize ship captured by a privateering expedition led by Captain Woodes Rogers and Selkirk's former boss, Captain Dampier. Alexander Selkirk returned to England and became a lieutenant in His Majesty's Navy.

A Reminder

On a clear day the green hills of the Juan Fernandez isles can be seen from twenty miles or more at sea. Atop a hill on Mas-a-Tierra, about 3,000 feet in the air, is the place Alexander Selkirk chose for a lookout post, scanning the sea day after day for a ship that might rescue him. One hundred and fifty-nine years after the castaway left Mas-a-Tierra forever, his lookout site was marked by a metal tablet set in rock and bearing this legend:

IN MEMORY OF
ALEXANDER SELKIRK
MARINER

A native of Largo in the County of Fife, Scotland, who lived on this island in complete solitude for four years and four months. He was landed from the Cinque Ports galley, 96 tons, 18 guns, A.D. 1704, and was taken off in the Duke, privateer, 12th February, 1709. He died Lieutenant of H.M.S. Weymouth, A.D. 1723, aged 47. This tablet is erected near Selkirk's lookout, by Commodore Powell and the officers of H.M.S. Topaze, A.D. 1868.

During his island exile Alexander Selkirk lived in a cave at the head of Mas-a-Tierra's bay. That cove subsequently was named Robinson Crusoe Bay.

16

Navy of the Lost

With a possible exception of certain yet-unexplained natural marine phenomena, the sea's supreme mysteries are vessels that have disappeared from the face of our globe without a trace. But those riddles have close rivals. They are the ships that have been found either with all hands strangely missing, or still aboard but done in, and no clues whatsoever to what happened.

Most publicized and longest perpetuated of these mysteries is the derelict *Mary Celeste*. There have been scores of others, taunting us over the centuries. And a distinct possibility exists that numerous other derelicts escaped discovery, even in the present era, transported into remote waters by currents and winds, cut off forever from traveled sea (and air) lanes. Who knows but what there may still be derelict vessels of fifty, even one hundred or more years of age that somehow have managed to resist time's ravages to continue their aimless wanderings in far-off corners of the oceanic world.

Unhappily, many discovered derelicts have been allowed to drift deeper and deeper into obscurity; and as the mists of time close about them, only scanty details come through to us. If only they had command d the attention and historical publicity of the *Mary Celeste*!

A Long-Dead Drifter

The *Marlborough* was a three-masted Scottish schooner. She set sail from New Zealand on a cargo-laden return to England early in 1890. A Captain J. Hird commanded her crew of twenty-three. She also carried an unspecified number of passengers on the voyage. The *Marlborough* was last reported in the Strait of Magellan, that winding, 370-mile passage at South America's lowermost tip, between the continent's mainland and Tierra del Fuego archipelago. That report described her as making normal progress.

But the *Marlborough* never reached England, and a search in the spring of that year failed to find her. She was written off as lost with all hands, not an uncommon occurrence in the sailing ship era, especially in the forbidding, stormy Cape Horn region.

Twenty-three years later, in 1913, the *Marlborough* was discovered, a roving derelict with skeletons for a crew and passengers. One could suspect that a plague killed them all, but what really happened was never established.

A Frosty Secret

I've always subscribed to a romantic notion that certain derelicts have strayed into polar regions to become locked in ice, their people preserved by the cold in their various attitudes when death overtook them. Perhaps imprisoned in Arctic and Antarctic ice is a veritable museum of vessels dating back to man's earliest sea ventures. That speculation has intrigued me for a long time, and it was fueled several years ago by the finding of the remains of a young mastadon, uncountable thousands of years old, preserved in ice.

Some substance was also given to this theory by the strange story of the English schooner *Jenny*, a story that first came to light in 1860.

In September of that year the *Jenny* was found derelict in the region of Drake Passage, a strait near Cape Horn connecting the South Pacific and South Atlantic. When the whaler *Hope*, commanded by a Captain Brighton, came across her, the schooner's officers and men were all dead but had been preserved by polar cold. From the *Jenny*'s log and her position when found, a fascinating story was pieced together.

She had sailed from Lima, Peru, during the late months of 1822. A little over a year later, in January of 1823, she was trapped in Antarctic ice. A final entry had been made in the ship's logbook on May 4, 1823, by her master: "No food for 71 days. I am the only one left alive."

Her icy prison had finally released *Jenny* after holding her for more than thirty-five years.

Another Chiller

For this one we have to return to the year 1775.

The whaler *Herald*, out of Greenland, was operating in polar waters west of that giant island. After a becalming in a forbidding place amid the menace of icebergs and floes, she moved into a dramatic weather change. Winds rose to gale force, tearing at her rigging and spars. A blizzard reduced visibility to near zero. Shuddering thuds and spasmodic hesitations of the vessel kept everyone aboard mindful of excellent chances of being stove in by ice. All through that wicked night the *Herald* made her uncertain way. Toward morning the storm abated, and dawn promised fair weather.

The sun was barely up when a lookout aloft shouted to announce a ship off to starboard. The stranger's appearance was sudden, but she hadn't materialized out of nothing. She had been in the area all the time, but hidden by an iceberg. At first glance this sailing vessel could have been carved from ice. The glistening stuff sheathed her hull, deck, masts, spars, shrouds and ratlines. The *Herald*'s master, Captain Warren, studied her through his glass as she came closer, gliding along a course that would take her within a few hundred yards. His sweeping lenses could pick up no signs of life aboard.

Nor did any appear as the eerie stranger, her coating of ice sparkling like diamonds in the morning sun, drew abeam of the whaler. Captain Warren hailed the ship. No response. When repeated calls failed to elicit a reply, he ordered a smallboat launched. He and a small party rowed toward the ice-coated hulk. On her stern they could just about make out a name: *Octavius*. On the face of the situation, a speculation was that this derelict had been locked in ice somewhere, then released by a thaw to drift.

Better Left Alone

Against their better judgment, but compelled by irresistible curiosity, Captain Warren and his men boarded the *Octavius*. No member of her company, alive or dead, was on deck. The deathlike silence aboard was broken only by the usual creaking of a ship's timbers and the crisp crunch of ice underfoot as the boarding party investigated topside. No corpses were under the drifted snow either.

Now they had to go below, a chore, we can be certain, the boarding party was loath to do. Wind-drifted snow was banked against the door to the companionway leading below to the crew's quarters in forecastle, and ice had sealed it shut. Finally they forced it open and gingerly went below—into a morgue. Preserved by the fierce cold were between twenty-five and thirty men, just as they had died, bundled in clothing and blankets in a pitiful effort to ward off the inevitable.

Another bad dream awaited them in quarters aft. In his cabin, the skipper of the condemned ship had frozen to death while seated at a

crew back on board for resumption of the run to Norfolk. On January 9, 1921, she departed Barbados, without cargo.

Two weeks later, on January 23rd, she was sighted by a lightship at Cape Fear, North Carolina, moving briskly northward. Nothing unusual was noted by lightship observers as the schooner passed.

Within the next couple of days a vicious nor'easter roared in to smack a large segment of coast, its gale-force winds whipping the Atlantic into mountainous seas. The *Carroll A. Deering* survived this big blow, and after it subsided was next sighted on January 29th by another lightship, that at Cape Lookout, off North Carolina's dreaded Diamond Shoals. Nearly a week had elapsed between the two lightship sightings, but she had covered less than 100 miles. Somewhere along the line she probably anchored to ride out the nor'easter.

This time things did not seem right on the schooner. Most noticeable, members of her crew were gathered on the quarterdeck, an assemblage that wouldn't have occurred if all were well with the captain. The irregularity hinted strongly at a lack of authority and discipline, and becomes graver in light of the crew's previous unruliness. As the five-master passed, one of her men called to the lightship through a megaphone, shouting that they had lost two anchors and asking that this message be relayed to shore. It may have been the last word from anyone aboard the *Carroll A. Deering*. Relaying the message to shore was impossible at the moment because the lightship's radio was temporarily out of commission.*

Next comes a strange interlude. Not long after the *Carroll A. Deering* passed the Cape Lookout lightship she was followed by an unidentified steamer, also northbound and on the same course. This second ship not only failed to identify herself (at that distance her name couldn't be made out), but also ignored the lightship's request that she pause long enough to take an important message. The lightship wanted to relay the *Carroll A. Deering*'s report of two lost anchors. The unidentified steamer continued her way up the coast. Later it was thought that she might have been the *Hewitt*, which subsequently vanished without a trace, Bermuda Triangle style.

Now, the Mystery

The *Carroll A. Deering* was hard aground on treacherous Diamond Shoals, where she was spotted early in the morning of January 31st by a man on watch at the local Coast Guard station. All sails except one jib were set, but binoculars failed to see any

*Accounts make no mention of radio aboard the *Carroll A. Deering*, which leaves us with the assumption that she didn't have one.

crewmen in evidence. Attempts to board her were thwarted the next day or two by ugly seas. Meanwhile, the U.S. Coast Guard cutters *Seminole* and *Manning*, along with the wrecking steamer *Rescue*, arrived on the scene. When the seas subsided, an agent of the Merritt-Chapman salvage company managed to board the stranded schooner.

As mentioned, all sails except one jib were set. The five-master apparently had been on a northeast heading, close-hauled on a starboard tack. Not a single human being was on board; and no attempt had been made to free her from the shoals' grip, a maneuver well within possibility since the ship carried no cargo at the time. Such apathy concerning the magnificent *Carroll A. Deering*'s grounding was extremely perplexing. A master wouldn't be likely to abandon his ship until every effort had been made to save her. Even then, there would have to be some extremely serious, imminent threat to prompt him to order abandonment. Still further, the ship was reasonably near the coast, and Captain Wormell must have known about the Coast Guard station.

Those peculiarities, coupled with that earlier contraregulations gathering of the crew on the quarterdeck, hinted darkly that something grim had happened to her skipper. Did he fall ill and die? Had he been murdered? Did he go over the side, accidentally or otherwise? (Perhaps there was a clue in those cryptic passages in his letter from Barbados.) Or had something happened to him during that violent nor'easter—a fatal injury, perhaps?

Another puzzler: Captain Wormell had marked his ship's course on a coastwise chart up until January 23rd, at which time they were off Cape Fear. Then someone else, unidentified, took over and continued marking the route until about seven days before the schooner was first seen aground on Diamond Shoals. Also very strange, Captain Wormell's luggage was gone. So were the vessel's papers, logbook, compass and chronometer. Two cats represented the only life aboard. In the only pleasant note in the entire affair, they were saved and given a new home on the *Rescue.*

The *Carroll A. Deering* had worked her way deep into the sand of Diamond Shoals. Water was just a few feet below deck level. Any traces of bloodshed below would have been washed away. There was no evidence of collision; but her steering equipment, including rudder and wheel, as well as her binnacle or compass housing, had been smashed. A huge sledge hammer lay nearby. Foc's'le and cabins had been cleared of personal belongings. Provisions were in the galley, and an evening meal had been in preparation during the crew's final moments aboard.

Indications pointed to the *Carroll A. Deering* being abandoned sometime between the afternoon of January 29th and the early morning hours of January 31st when she was sighted hard aground. The schooner's crew was experienced, so it was unlikely that they would have left their ship, especially with any heavy seas running, just because she was aground. The *Carroll A. Deering* had not started to break up. Yet all her men were gone, and two of her boats were missing. No rope ladder dangled over the side, and the boats' falls had been cut. Deepening the mystery, two red lights had been placed high in the rigging, the signal indicating a vessel aground, abandoned at sea, or out of control.

The disappearance of all hands from the *Carroll A. Deering* has yet to be explained satisfactorily.

A Fleet of Conjectures

Considering Captain Wormell's known problems with his first mate and crew, mutiny had to be considered. But what could have caused it? Captain Wormell was a religious man. Had he wearied of the crew's drinking and become overly strict, whereupon they rebelled? Or, on another conjecturing tack, had they forced him to abandon ship, either during the northeast storm or when the vessel ran aground? It didn't sound plausible that a seasoned crew would panic under such circumstances.

And what of Captain Wormell? Was he murdered, or thrown overboard, whereupon the crew fled their ship in fear of reprisal? If the captain were ill or had met with an accident by the time the *Carroll A. Deering* passed the Cape Fear lightship, it was most extraordinary that *that* fact—rather than the loss of two anchors— wasn't reported. And while on that angle, why had a crewman, rather than the master, announced the anchors' loss? One thing seemed certain: If the crew did indeed abandon the vessel, for whatever reason, their two boats probably were swamped by turbulent seas.

All possibilities were considered. Thought turned to the idea that the unidentified steamer—later thought by some to be the missing *Hewitt*—picked up the *Carroll A. Deering*'s men, after which the rescued met the same fate as their rescuers. The complete disappearance of the *Hewitt* is another of the sea's unsolved mysteries.

The United States was thirsting in a nationwide drought imposed by the Volstead Act—Prohibition—at the time, and smuggling in illicit liquor was a lively, highly profitable business along many segments of the Atlantic seaboard. Considering Captain Wormell's religious nature and impeccable reputation, it's extremely unlikely that the *Carroll A. Deering* would bring a cargo of liquor from Barbados. However, theorists suggested that possibly there was

125

collusion between members of the schooner's crew and smugglers, planned in a waterfront tavern, to take over the vessel. Flaws in that theory were that she was just too big, too slow, and, as a five-masted sailing ship, much too conspicuous for such illegal business. Besides, the conspirators would have had to be out of their minds to attempt such a foolhardy stunt so close to the mainland coast.

About three months after the *Carrol A. Deering* went aground, a man named Gray reported finding a message-bearing bottle washed up on shore while strolling along a North Carolina beach. The message, he said, mentioned the schooner being captured by an "oil-burning" boat, "something like a chaser" (presumably meaning similar to a Coast Guard vessel employed in chasing rumrunners) that was "taking off everything." Members of the schooner's crew were trying to hide, the note continued, but were being removed in handcuffs. The bottle message concluded with an appeal to notify the *Carroll A. Deering*'s owners.

This note became a mystery within the mystery. Handwriting experts studied the missive and traced its distinctive longhand to Herbert Bates, an engineer on the ship and a loyal crewman trusted implicitly by Captain Wormell. At least, its handwriting was uncannily identical to Bates's, and a forgery under the circumstances would have been impossible. Bates, it was agreed, might have had an opportunity to scrawl the note while still hiding below deck.

Meanwhile, U.S. Government agencies had been drawn into the investigation, not only because of the mystery of the *Carroll A. Deering*'s crew in territorial waters, but also because of strange disappearances of other vessels off the coast. Some government investigators subsequently declared in effect that the message in the bottle was a hoax, but failed to explain how anyone could have duplicated Herbert Bates's handwriting closely enough to deceive not one but three calligraphy experts. Presumably the finder of the note didn't even know Bates. In counterargument, some observers theorized that maybe Mr. Gray had been coerced into confessing a hoax rather than be drawn further into the investigation. This mystery within a mystery was either never solved or a solution was never publicly announced.

Pirates?

As we ordinarily think of piracy, its existence in the 20th century is hard to believe . . . until you consider that we still have a form of it, but under a different name: Hijacking. But piracy or hijacking wasn't logical in the case of the *Carroll A. Deering* because she purportedly was running to Norfolk without a cargo, a fact undoubtedly known on the Barbados waterfront, where pirates or hijackers would have had spies and informers. The hijacking theory would have had substance if the vessel had been known to be carrying liquor.

Nevertheless, that controversial note in a bottle led to a speculation that the *Carroll A. Deering* had been overtaken by Russian pirates who had kidnapped her crew to Vladivostok. When and where this could have happened was not made clear. Nor is it clear why pirates or hijackers would bother with the crew and not the vessel, unless some attempt was made to seize the ship, which failed (maybe when she ran aground), whereupon her crew were taken prisoner to eliminate witnesses.

For several weeks after the *Carroll A. Deering*'s downfall, a number of vessels vanished without traces off the U.S. East Coast. They included the steamer *Hewitt* and some others of fair size. Newspapers mentioned between eight and a dozen such mysterious disappearances. During the first few vanishings it was thought that the vessels simply were victims of one type of marine disaster or another; but as the number mounted within a relatively short time the situation looked very peculiar, enough so for the U.S. Government to investigate.

Along about that time the Russian pirates theory seemingly was given substance by a newspaper report that a number of vessels, thought to be foreign but with their names obliterated, were seen being brought into Vladivostok, manned by Russian crews. However, relations between the United States and Russia were anything but cordial, so the piracy angle came to nothing. It wasn't very logical anyway. Why would Russian pirates or hijackers operate in U.S. coastal waters, so far from their supposed bases of operation? It would have been extremely risky. Then, as now, the Atlantic Coast was trafficked heavily by vessels of many kinds, and there were Coast Guard installations up and down the seaboard. Besides, it would have been easier and more practical to prey upon shipping nearer their bases of operation.

And Omega

Whatever really happened aboard the *Carroll A. Deering* seems most likely to remain a big question-mark forever.

There's no mystery in her exit from this world. Handsome and only a few years old—much too young to die—she became another victim of Diamond Shoals. For a long time she remained more or less intact, graphic testimony to her construction; but she had settled so deeply into the Diamond Shoals' vise that she couldn't be refloated and freed. Anything worth salvaging was removed, and her stripped body was abandoned again, this time to the not-so-

127

tender mercies of the sea and elements. Even so, she continued to linger in slow disintegration. Finally there were plans to scatter her remains by blasting. A sudden, severe storm eliminated the need. When the storm abated, the once-proud five-master had been reduced to driftwood.

If the *Carroll A. Deering* knew the story of her master and crew, she took it into eternity with her.

17

Fire Island Tales

Along the southern coast of Long Island, New York, extends a chain of long, narrow, sandy islands, separated by inlets from the Atlantic Ocean. Technically they're known as barrier beaches. Anyone wanting to be insulting could call them overgrown sand bars—which is what they are, in a way. They were created by a build-up of sand, unceasingly chewed from Long Island's eastern reaches by a restless ocean, and then transported westward by waves and currents to be redistributed elsewhere along the main island's coast. This is a natural process that has been going on continuously for uncountable thousands of years.

The sandy isles are termed barrier beaches because they extend as a bulwark against the ocean. In this they provide a double bonus, for by insinuating themselves between the Atlantic and the Long Island mainland they also create a great chain of bays which provide swimming, boating and fishing pleasures for thousands of Long Islanders and their visitors, as well as a livelihood for generations of clam diggers.

Two of these barrier isles are known throughout the United States, at least by name. One is nationally famous because it contains Jones Beach State Park, a seaside playground which has been visited by folks from every state in the Union, to say nothing of people from many foreign countries.

The second barrier beach is Fire Island. Throughout the 1930s Fire Island was a quiet, isolated place, accessible only by boat. It had only a few established summer bungalow colonies, and other visitors consisted of picnickers and surf fishermen coming over from the mainland by ferry or private boat. Most of Fire Island still can be reached only by water, but since World War II it has become widely known as one big summer resort as increasing numbers of sun-seekers "discovered" it. Cottage colonies have sprouted all over the place. Trains bring hordes of weekend visitors from the metropolitan area to waiting ferries. Thousands of others enjoy Robert Moses State Park, a seaside playground on the island's westernmost end, linked with the mainland by a causeway. Except for that causeway, there are no roads on Fire Island. Eventually there will be, I suppose. Meanwhile, transportation between summer colonies will continue in the form of beach taxis, scuttling back and forth along the oceanfront strand.

Still, there are parts of the isle, like Fire Island National Seashore, that retain their pristine charm. Away from lively summer activities, there are places where the scene is still much as it was a hundred or a thousand years ago. There are stretches of wide, empty beach, flanked by lonely dunes, where the wind sighs through wild grass, where the surf varies its tempo between soft, lazy "swooshes" and a thunderous roar of breakers, and where often the serene quiet is fractured only by the raucous cries of gulls. Such scenes help to perpetuate the island's fascinating history.

How Fire Island Got Its Name

Long Island's Indians knew the place long before the first English settlers infiltrated the region during the first half of the 1600s. But it remained for the Fire Island name to be bestowed by the newcomers.

There are two schools of historic thought as to how the name came about. In one, the word "fire" is believed to have been prompted by fires built along the island's beach by early whalers. During early decades of Long Island colonization, English settlers learned about shore whaling from the Indians and turned it to commercial advantage. There were enough whales in those days for many of the beasts to come fairly close to shore, close enough for the Indians to attack and kill them for parts that were put to various tribal uses, chiefly religious. The settlers soon realized the commercial possibilities of whale oil and developed a type of surf-launched boat from which the huge mammals could be harpooned.

After killing, the animals' blubber or thick skins were stripped from the carcasses, cut into chunks, and "tried" or rendered for oil, desirable as a fuel and a lubricant. "Trying" was accomplished in large kettles on shore, and the liberated oil was shipped to Dutch merchants in an upcoming town called Nieuw Amsterdam, way down to the westward. A theory is that the "Fire" in "Fire Island" stemmed from fires under the try kettles.

The other theory concerning Fire Island's name is based on a supposed misunderstanding. This school of thought believes that there once existed a cluster of five islands in that area, which were so labeled—*Five* Islands—on early charts. On old charts the letter "v" often was printed in such a way that it could be misread as an "r."

No matter. One way or the other, the name Fire Island stuck.

Land Pirates

Actually, there's a third, sinister explanation of the word "Fire" in the name. Years ago this explanation was disputed vigorously.

In a distant epoch, land pirates, also known as "wreckers," carried on a nefarious practice along numerous segments of the U.S. Atlantic seaboard. Theirs was a simple but quite effective *modus operandi*. On dark nights in remote, lonely places the rascals built fires or beacons, or waved lanterns that could be seen at sea, to lure unsuspecting vessels into thinking they were being guided into an inlet or a harbor. What they were being guided to, of course, was a sand bar or shoal where the vessels would run aground or be wrecked, whereupon the shoreside pirates stripped them of everything of value. Presumably the looters were prepared to handle any resistance.

As late as the 1800s reports of such shenanigans gave Fire Island—and eastern Long Island in general—a bad name among mariners. New Jersey didn't exactly have a lily-white reputation either. In an article published in 1876 an old Barnegat fisherman talked about land piracy, pointing out that there were niceties even in that profession and leveling some pretty stiff charges at Long Island. Whatever went on along the New Jersey coast was a church social in comparison with the skullduggery on Long Island, he sniffed. "No man or woman was ever robbed on this beach [in New Jersey] unless they was dead," he declared righteously. "Of course," he quickly added, "I don't mean their trunks and such. I'm talkin' about their bodies." He went on to charge that Long Island "land pirates" cut fingers off living people for their rings,"..."but the Barnegat men never touch the body till it's dead, no sir."

Although historians have admitted that possibly land piracy existed to a certain extent on Fire Island and elsewhere on Long Island's South Shore at one time long ago, they have also contested it with very solid points. To begin with, for decades, enough vessels came ashore, ran aground or were otherwise wrecked of their own accord, spewing their cargoes all over the place, that land piracy really wasn't necessary. Looters merely had to be patient, and sooner or later some confused skipper would oblige. Furthermore, it just wasn't in the nature of local citizens to go in for land piracy. Of course, that doesn't take into consideration interlopers or bad apples in the barrel, but let that pass. Besides, if the truth were known, otherwise law-abiding residents probably swiped more than professional looters could.

131

Shore whaling remained a profitable industry on Long Island until the big marine mammals became both smarter and fewer and it was necessary to hunt them at sea in ships, sometimes for as long as three or four years to get a good payload of oil. Fire Island was a remote, very lonely place in shore whaling days. Aside from sea birds soaring and swooping and little fiddler crabs scurrying along the bay shores, the only signs of life along that windswept barrier beach were the scattered whaling crews who lived among the dunes. Crude huts sheltered them from the elements. With their whaleboats ready for instant launching in the surf, the crews spent their days peering out to sea and listening for their lookouts' cry of "Whale off!"

In that distant era Indians often were enlisted for whaleboat crews, as harpooners and as wielders of the lance that finished off the animals, because of their strength, stamina, good eyesight and ages-old knowledge of the big beasts. Half of the members of one Fire Island crew were Shinnecock Indians, and this particular squad of whaleboatmen resided in a primitive, weathered structure on the beach near a place called Whale House Point.

Day in and day out they tenanted the beach, lest they miss a passing whale. Their complete isolation necessitated periodic trips to the Long Island mainland for provisions; and so that their daytime vigil wouldn't be interrupted, the ferrying of supplies was done at night. Since it might be dark when the boats returned from the mainland, a lookout with good hearing and night vision was posted to watch for them and guide them with a beacon fire on the sand beside Great South Bay.

Enter Jonas

On the night of our story a man named Jonas drew the boring assignment of watching for the supply boats. In such duty the hours of darkness crept by at a snail's pace; on this particular night they seemed to stand still. Finally boredom overcame Jonas, and when he felt that no boats would be returning for quite some time, he decided to stroll eastward across the dunes. At least it was something to do.

Besides, he was an inquisitive fellow, this man Jonas. While on watch the previous night he had had an eerie experience. Although a vague foreboding advised him against it, he wanted to see if the occurrence would be repeated. He set out across the rises and hollows of the sand hills.

Atop an especially high dune, the place he had visited the night before, he paused and cocked an ear, listening very carefully for a repetition of some weird sounds. For many minutes he heard nothing extraordinary, only the wind's soft whisper and the breakers' restless rumbling on the oceanfront beyond.

Then his keen ears detected it.

There was no mistaking that sound. From a nearby hollow among the sand hills came a moaning rising in the darkness. This was no freakish noise made by the wind, Jonas knew. The wind wasn't strong enough, and there was no structure thereabouts for it to caress to generate such a sound. It was a moaning that could have been human, yet it had a definitely supernatural quality . . . and Jonas knew he was quite alone in that place.

When he first heard it, the moaning was soft and low. As he listened, it grew louder, then louder yet, and finally rose to a shrill, blood-chilling wail. It wafted away across the tops of the dunes, as though riding the night wind. Soon it was a distance away, but still audible. Off in the darkness the wail gained volume, as though returning, then it ceased abruptly. Jonas felt the hair on the nape of his neck rising, and a sudden chill made him shiver. He hurried away from that haunted spot. When a check at his lookout post convinced him that the supply boats would not be returning until after daybreak, he threw himself into a bunk in the crew's shack. The first pale fingers of dawn were clutching at the horizon before he dozed off in a troubled sleep.

"Whale Off!"

Later that day Jonas and his companions had the good fortune to sight a whale close inshore. Swiftly a boat was launched in the surf. Strong backs bending to the oars, they made for a area where they calculated the giant animal would surface to breathe. The breaching would be signaled by the appearance of the beast's broad back, and a spouting, sounding like a thunder clap, as the whale exhaled. The harpooner was already in the bow, his long weapon poised.

Their calculations were accurate. A monstrous dark back parted the sea nearby, and the whale's noisy exhalation shot a geyser of vapor skyward.* The harpooner was about to fling his iron when the animal unexpectedly moved closer, forcing the boat to maneuver quickly or risk being smashed to kindling when the whale's big tail flukes slapped the surface. Before they could return to a position for

---

*Whale hunters can identify different species of whales by the height and nature of their spouts.

flinging a harpoon—irons were hand-thrown in those days—the behemoth perversely started to swim away, out of range. In desperation the Indian harpooner let fly. It found its target, and its "lily" or spearhead sank deep into the animal.

Seconds later, the whale turned and charged the boat, striking her bow with its massive head. This was no time to assay any damage. All hands dove overboard, and came to the surface just in time to watch the infuriated giant demolish their boat. Luckily, a breeze was blowing briskly toward the beach, so assisting the crew to reach shore. All of them, that is, except the harpooner. He had vanished, never to be seen again.

Jonas said nothing to his companions about the haunting moans and wails he had heard among the dunes on two nights before the tragedy, but he asked himself if there was some kind of supernatural connection between those occurrences? Jonas couldn't find an answer, but his curiosity had reached its limit. He never remained overnight on that beach again.

Later, others reported that for several nights prior to the harpooner's drowning *they* had heard ghostly calls and shouts among the lonely dunes. It was as though men were calling to each other; but no one was there, and the voices weren't like those of living humans. An old man who lived on the beach to shoot wildfowl said that twice in his life he heard moaning in the dune hollows, and that afterward each time he found bodies on the beach.

But the harpooner's corpse didn't wash ashore.

One Big Cemetery

Several locations along North America's eastern seaboard have been dubbed "Graveyard of the Atlantic" at one time or another. Among the most infamous are North Carolina's Outer Banks and Diamond Shoals. So many vessels have met doom in this area that an accurate count is virtually impossible.

Diamond Shoals often has been termed the worst ship graveyard of the lot, which may or may not be true. If it is, the area has a close contender for the dubious distinction in the southern coast of Long Island. More than 300 years' worth of sea history has been written along that coast. Even the late Jeanette Edwards Rattary, newspaper publisher and Long Island's greatest maritime historian, probably would have hesitated to venture an estimate of how many vessels of assorted types—sloops and schooners to motorized fishing boats to liners and warships—have encountered calamity in the island's waters; and she chronicled such events for many years, accumulating a list of hundreds.

Fire Island is part of that maritime cemetery, and over the years has contributed a quota to the total. One of the first casualties of which there is record, up at the very top of Mrs. Rattray's long list, was the *Prins Maurits*. The date of her demise at Fire Island was the eighth or ninth of March, 1657.

The *Prins Maurits*—*Prince Maurice*, as you've guessed—was named for Maurice of Nassau, son of William the Silent and Prince of Orange, a Dutch military hero. In December of 1656 she sailed from the Netherlands on a voyage to the New World. Aboard was a human cargo that totaled about 130, most of whom were colonists looking ahead to a new life far across the briny.

No specifications of the *Prins Maurits* are available, but she probably was a comparatively small sailing ship, considering her era. Certainly she was a small vessel for crossing an ocean notoriously harsh in its northern expanses in winter. She must have encountered storms, and we can't imagine what agonies of fear, uncertainty and seasickness her people must have suffered. But she did weather the crossing and in early March was approaching Nieuw Amsterdam, a port better known today as New York.

Late on a March night the valiant vessel went aground on Fire Island, then called South Beach. Weather was severe: Bitter cold, with drifting ice added to sand bars as a hazard. Wretched weather continued the next day, but somehow the ship's boats made their way through rough seas and chunks of floating ice, then through a violent surf, to reach shore safely. There is no better tribute to Dutch seamanship than the fact that no one was lost.

On the beach the people from the wrecked *Prins Maurits* suffered in the freezing, biting cold, for the place was barren of anything with which to build fires. Nor was there any shelter against the elements. During the next two or three days, however, the ocean shoved the stricken ship closer to shore, and men managed to salvage a quantity of articles before the *Prins Maurits* disappeared for all time. The castaways also encountered some friendly Indians who informed them where they were (how the two groups communicated isn't recorded). Subsequently they made their way to Nieuw Amsterdam, where they presumably lived happily as part of that Dutch community that was to become New York City.

Poor "Elizabeth"!

A Jonah or jinx sailed with the bark *Elizabeth* on her last voyage, and mariners of the old school swore it was because women were among her twenty-three passengers. Even as a source of passenger revenue, women were disliked by superstitious seafarers. They were bad luck aboard ship, that's all there was to it.

Be that as it may, the *Elizabeth* weighed anchor in the harbor of

Leghorn, Italy, for New York on May 17, 1850. Only two days out, her master took to his berth with smallpox, an even more serious illness in those days than now. Understandably, within the confines of a ship a contagious disease was greatly dreaded. Fears of a shipboard epidemic intensified when the captain died. And now, well out on the Atlantic, there was a double worry. Command had fallen to an inexperienced mate.

The extent of that character's seamanship was demonstrated with stunning impact later as the bark beat her way on a westerly heading off the southern shore of Long Island. At that point the mate thought he was off New Jersey. How he could figure that New Jersey was on his starboard side when heading westward is amazing in itself. But that's what he believed before he lost his bearings completely. In the resulting confusion, complicated by a strong wind, he converted the ship into a pile of junk on a Fire Island sand bar at four o'clock in the morning—a dalliance, you can be sure, that did little for passengers' sleep.

Although the *Elizabeth* stayed afloat for many hours and lay only fifty yards off the beach, no one dared launch lifeboats in the ugly sea then running. Some of the more desperate passengers boldly lashed themselves to timbers and managed to reach shore safely, but the majority remained aboard and awaited providential developments. This also required considerable courage, since the bark was in imminent danger of breaking up and dumping them all into the ocean.

One of those taking their chances aboard was Margaret Fuller, noted writer, women's suffragist, and former literary editor of the *New York Tribune*. Margaret sat on deck, hunched against the main. t, her hair streaming wildly in the wind. Clutched in her arms w. her young son, bundled in shawls. With wind shrieking a dirge in the rigging, her situation took a bitterly ironic turn. A ship's steward, seeing that the mainmast might come crashing down at any minute, and knowing Mrs. Fuller's emphatic refusal to leave the *Elizabeth*, he seized her child and leaped overboard in an attempt to reach shore. Both he and the boy drowned. For the youngster it was an especially ironic end, since he had survived a severe attack of the disease that felled the captain.

Not long afterward, Margaret Fuller herself was swept away and carried out to sea. Her body was never recovered. This too was a special slice of irony, for she had risked her life in a suffrage cause in Italy. The Atlantic seemed determined that not a single trace of this poor woman be left. A monument erected to her memory at Point O' Woods, Fire Island, in 1901 was also claimed by the sea.

The *Elizabeth* tragedy cost nine lives. Twelve people were saved.

Then came an ugly postscript. A newspaper charged that "land sharks and pirates" made off with nearly everything washed ashore from the *Elizabeth*—not only cargo, but also passengers' clothing and personal belongings. Some forty people were found guilty of capitalizing on the wreck. In one home alone, it was reported, one thousand pounds of silk were found.

18

Holocaust on the Mississippi

Prior to development of nationwide networks of railroads and highways, the major thoroughfares of midland America were the big rivers. And in this system the mighty Mississippi was Main Street. Arising in Lake Itasca in northern Minnesota as a modest stream, it ambles 2,470 miles* southeastward, then southward, widening and deepening as it goes, and finally exits into the Gulf of Mexico through its fan-shaped delta below New Orleans.

Up and down this enormous waterway paraded all kinds of river vessels, from the skiffs of Huckleberry Finn's epoch, to homely, freight-totin' keelboats, to majestic passenger packets such as the *Robert E. Lee*, *Natchez*, *Pennsylvania*, and many others. With their paddlewheels churning the muddy Mississippi into a brown froth, these passenger packets were the big-river equivalents to the ocean liners that followed decades later.

The *Sultana* was such a packet, but her life had a far more terrible finale than that of most of her sister steamers. The *Sultana* disaster is relatively unknown in comparison with the sinking of the *Titanic* and *Lusitania*, yet its loss of life was as great. In fact, no marine disaster involving American ships comes anywhere near matching the *Sultana* horror in that department.

The 1,720-ton *Sultana* was of typical Mississippi River packet design: Three decks crowned by a pilothouse; private accommodations for passengers, with public rooms for dining and socializing; a roomy main deck on which to stow bales of cotton and other cargo; and huge side-paddlewheels turned by a steam engine. Often the first glimpse of these packets as they rounded a bend in the river was the towering twin stacks, slender and black, belching smoke into the Southern sky.

The *Sultana* was designed to carry about 400 passengers in addition to a good payload of cargo. According to a record of her which I have, she was a young vessel at the time of her demise, only two or three years old, having been built in Cincinnati in 1863.

*If we include its largest branch, the Missouri River, the Mississippi system totals approximately 3,872 miles.

There's no information about how well she was built; but it's important to note that river steamboat companies' competition for revenue was so fierce at that particular time that construction often was hasty and poor. For the same reason, maintenance wasn't always what it should have been; and there was a lack of proper safety codes. Disastrous boiler explosions and fires destroyed a number of the packets before they could give ten years of service. Such an explosion on the *Pennsylvania* killed Henry Clemens, Mark Twain's younger brother.

Her Date with Death

The long, bloody Civil War was over. In the warmth of an April day in 1865 a large contingent of victorious Union troops massed on wharf in Vicksburg, Mississippi, awaiting steamer transportation to St. Louis, first leg of a long-dreamed-of return to their homes in the North. Among them were many wounded. Numerous others were ill or emaciated after a year or more in overcrowded Confederate prisoner-of-war camps. The steamer they awaited was the *Sultana*.

Even in their weakened condition the soldiers' elation surged as they watched the *Sultana* warp in alongside the wharf. Chances are, though, that their elation didn't last long when they saw how they would be herded aboard like cattle. It was never determined exactly how many men were put aboard the vessel that day, but history pegs the figure at somewhere between 2,000 and 2,500—this on a packet designed to carry 400. Also taken aboard were unspecified numbers of horses and army mules, much military gear, and a miscellaneous cargo that included one hundred heavy hogsheads of sugar.

Perhaps it was in protest that her steam whistle became stuck in a piercing shriek as the dangerously overloaded sidewheeler groaned away from the wharf. Until it could be turned off, the whistle's piercing shrillness was like a file rubbing on the already raw nerves of the soldiers.

That night, as she threaded her way through small islands about two hours upriver from Memphis, the *Sultana* was torn apart by a tremendous blast. Her boiler pressure had been forced far above a safe maximum to keep the overloaded vessel under way against the Mississippi's current. The explosion blew out bulkheads, shot hatch covers skyward, and tore up sections of deck, catapulting passengers in all directions. Escaping live steam scalded men and animals, and their screams mingled with frantic yells. In the ensuing panic, ill and injured soldiers, too weak to get out of the way, were trampled to death by their fellows and stampeding beasts.

And that was only the first phase of the catastrophe.

The second phase followed the explosion swiftly. Flames mushroomed from the ruptured engine-room. Greedily feeding on wood everywhere, the fire climbed topside and raced forward and astern, quickly engulfing the *Sultana* and turning her into a crematorium. Many perished within minutes. Dozens leaped into the river, many with their clothing and hair on fire. Others jumped to a death by drowning. Consumed by the fire were the uncountable injured and dead from the explosion.

A reliable estimate is that approximately 1,500 soldiers lost their lives in the *Sultana* disaster. What an ironic end for men who had survived combat and the rigors of primitive prisoner-of-war camps!

When Albert Hicks "jigged on air" at Gibbet Island in New York Harbor for multiple homicide at sea he attained the dubious distinction of being the last man hanged for piratical murder in the United States. See Chapter Three, "The Hanging of Albert Hicks."

(Larry Kresek)

The two boys had been romping among the sand dunes. Then, in a lonely place, their play ceased abruptly as they came upon a sight that transfixed them. Thus did a skeleton introduce a fascinating tale. See Chapter Four, "Legend of the Money Ship." *(Larry Kresek)*

During her career the excursion steamer *General Slocum* provided escape and pleasure for many city folk. Then, suddenly, on a nightmarish day in June of 1904, she became a death trap and crematorium for hundreds of children. See Chapter Five, "Vessels From Hell." *(Peabody Museum of Salem)*

An old engraving pictures the brigantine *Mary Celeste* as she may have appeared when she was discovered derelict and adrift, all hands having vanished without a trace. From Chapter Six, "A Femme Fatale Named 'Mary Celeste'." *(Peabody Museum of Salem)*

For nearly half a century the ocean liner
Great Eastern was among her era's mar-
vels, the largest ship in the world. To
mariners she was also a wonder in another way. They wondered how one vessel could have so much bad luck and
survive. See Chapter Ten, "Biggest Jinx
of All Time?" *(Harry T. Peters Collection, Museum of the City of New York)*

In eerie silence a glistening derelict suddenly glided out from behind an iceberg. There was no one at her helm, or anywhere on deck. When a party boarded the *Octavius* they discovered several ghastly reasons why. See Chapter Sixteen, "Navy of the Lost."
(Larry Kresek)

In this 1919 photograph the handsome
five-masted schooner *Carroll A. Deering*
was about ready for launching. Only
two years later she was found hard
aground on Diamond Shoals, North
Carolina, with her crew mysteriously
missing. To this day no one knows what
happened to Captain Willis B. Wormell
and his men. See Chapter Sixteen,
"Navy of the Lost." *(Peabody Musuem
of Salem)*

Among the sea's most provocative enigmas are ships found wandering with all hands dead. In long-gone years plague sometimes was the villain, but we can wonder if there may have been other horrors that would still be beyond human understanding. See Chapter Sixteen, "Navy of the Lost."
(Larry Kresek)

A camera's lens captures the ghostly figure of a long-dead naval officer striding the deck of an historic United States warship. See Chapter Nineteen, "Seafaring Spooks." *(George Cook, Baltimore Sun)*

Everything about the Dutch freighter *Ourang Medan* was weird: All her crew dead on board; a fire mysteriously erupting as rescue vessels were about to tow her to port; and an explosion that sent her to the bottom with her riddle forever unsolved. See Chapter Twenty-Three, "Death From Outer Space?"
(Larry Kresek)

This has to be one of the most amazing photographs ever taken. Aboard the Cities Service tanker *Watertown* an officer snapped the ghostly faces (in circles) of two recently deceased shipmates. The apparitions appeared in the waves alongside the vessel after the two seamen were buried at sea, and were readily identifiable as the faces of Michael Meehan and James Courtney. See Chapter Nineteen, "Seafaring Spooks." *(Cities Service Company)*

The remains of the *Morro Castle* attracted thousands to Asbury Beach, New Jersey, in 1934. See Chapter Twenty-Seven, "Deadly Secret of the 'Morro Castle'." *(Wide World)*

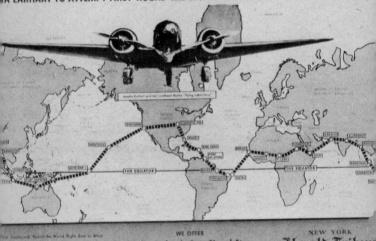

WE OFFER

Amelia Earhart's Own Signed Story

Her exclusive intimate keynote of this epic adventure sent to us by radio and cable from round-the-world aline *build-up* advance stories, exclusive photographs, and Miss Earhart's own handwritten record of the journey after its completion.

27,000 miles of flying by the world's foremost woman pilot

NEW YORK

Herald Tribune

Syndicate

The *New York Herald Tribune* ran this map of Amelia Earhart's proposed flight around the globe. Somewhere between Lae, New Guinea, and tiny Howland Island the plane and its intrepid crew vanished for all time. See Chapter Twenty-Eight, "What *Did* Happen to Amelia Earhart?" *(Lockheed Aircraft Corporation)*

Amelia Earhart posed with her Lockheed Electra, aboard which she and navigator Fred Noonan winged into oblivion in the Pacific Ocean's vastness on a July morning in 1937. See Chapter Twenty-Eight, "What *Did* Happen to Amelia Earhart?" *(Lockheed Aircraft Corporation)*

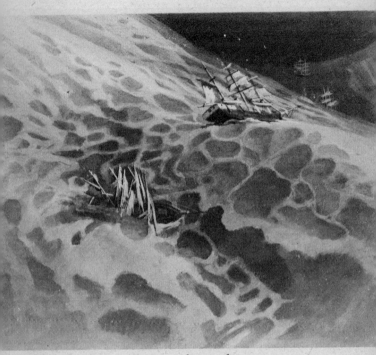

We probably will never know how many sailing ships survived Atlantic Ocean storms and mountainous waves, only to find their way later into that vast, becalming limbo called the Sargasso Sea. See Chapter Twenty-Nine, "A Legendary Sea of Doom." *(Larry Kresek)*

19

Seafaring Spooks

In the United States the Civil War was at its destructive zenith when the sailing ship *Usk* stood out from the Welsh port of Cardiff, laden with coal and iron, on a long trading voyage to Peru, 'way around on the other side of Cape Horn. But the Americans' internecine conflict was of no consequence to the British merchantman's master, Captain Richard Brown. He was more concerned with making up part of the time lost to foul weather on the voyage's transatlantic segment.

Storms astern of her for the time being, the *Usk* creaked along under bellying canvas toward that wild region of South America's Cape Horn. Rounding the Horn was something to look forward to with a certain amount of dread. It was a forbidding, tempestuous place, smitten by violent storms and often lethal to ships. Reaching shore might not be much of an improvement, since areas reportedly were inhabited by savage natives, not noted for their hospitality. Once the Horn was left far astern, and with decent breaks from weather, the voyage's Pacific segment could be routine.

A Warning and Indecision

Such, perhaps, were the musings of Captain Brown as he strode his deck one quiet, star-filled night. But whatever claimed his meditation, his train of thought was derailed rudely by a wraithlike figure of a woman appearing out of thin air in the shrouds. Sober, realistic masters usually were not given to seeing ghosts, especially any that pop up spontaneously. Captain Brown was suitably astonished, and before he had a chance to recover, the specter beckoned to him. Later he was to state that his legs seemed to function independently of his mind to obey the summons.

The phantom female descended from the rigging and stood before him. In a voice he could hear quite distinctly, the apparition warned him to put about at once and return to home port. Failure to do so, it counseled, would result in loss of the *Usk* and all hands. The vision smiled a farewell and evaporated as abruptly as it had appeared.

We can presume that Captain Brown was left with bulging eyes

141

and a sagging jaw. He also was left with a terrible choice. Should he heed the lady spook's warning or ingore it? He distinctly heard it, but from a supernatural figure. The easier choice, of course, would be to put the incident out of his mind as some kind of trick of the nerves—if he could, that is—and keep his mouth shut. No one would be the wiser, even if the prophecy came true. Besides, even if the warning became reality, the whole thing would be purely academic anyway, since the misty lady said all hands would be lost. On the other hand, supposing there was something to it? If he heeded the warning, how could he explain the reason to his officers and seamen? Or, what was even worse to ponder, how could he justify a money-losing voyage with the vessel's owners?

Captain Brown was still tortured by indecision when, soon after the encounter with the spectral woman, the *Usk* became enveloped in thick fog rolling in quickly from seaward. Whether it was intended to or not, this development helped Captain Brown to reach a decision. The heavy mist, he told himself, may have been to accent the warning. It looked like an evil omen, a harbinger of things much worse.

Warning Enough

Captain Brown summoned his first mate and some of the crew. While these hearties stared at him in disbelief, he announced with finality that the *Usk* was returning to Cardiff forthwith. He wanted no questions or discussions. But the first mate protested vigorously, pointing out a detail that Captain Brown realized all too strongly: The vessel's owners would be outraged at this radical alteration of plans. Unnerved as it was, Captain Brown flew into a rage when his decision was challenged, and cut the council short by ordering the first mate to be clapped in irons. The *Usk* put about and aimed her bowsprit toward Wales.

Scuttlebutt spreads rapidly within the confined community of any ship. It spread even faster aboard old sailing vessels. Somehow word of the captain's meeting with the spook got out, and soon all hands knew why they were headed home. The reason must have been received with mixed feelings. One camp—the firm believers in superstitions—had to be sympathetic to their skipper's decision, and grateful too. The other camp probably thought he was some kind of lunatic, spook or no spook, for risking his employers' wrath.

The *Usk* reached Cardiff six weeks later. As soon as she was tied up at a wharf, Captain Brown hurried to his company's offices to face the consequences. To say that the ship's owners were furious is an understatement. And their ire was understandable. For months the *Usk* had been at sea, consuming time—a valuable commodity in a competitive field—incurring expenses, and not earning a shilling. To top if off, she had returned with her cargo undelivered.

There was a formal inquiry. When all testimony had been heard, the wretched captain stood divested not only of his command of the *Usk* but also of his master's papers. Ridiculed, disgraced in his profession, he was a broken man and faded into obscurity.

Persistence

As soon as possible another master was engaged to command the *Usk*, still laden with the same cargo, to have another try at Peru. Any elation the new skipper felt at his appointment was immediately offset by problems. Word of the phantom lady and the pointless voyage had preceded him, with the result that he experienced great difficulty in enlisting enough men for a crew. Sailors were a highly superstitious band in those days. Whether the ghost in the rigging was only fo'c'sle gossip or otherwise, they were having none of the *Usk*. It took a bit of doing, but eventually the new captain succeeded in mustering enough hands. Again the ship's bellyful of coal and iron was outward bound for South America.

An unnaturally long silence followed. Then, about four months after the *Usk* left Cardiff below the horizon, a communiqué reached her company's offices, relayed by another vessel. The *Usk* had been destroyed by fire at sea, read the message, and all hands were lost.

Some Ghosts Wear Well

Often when phantom seafarers are seen they are manning spectral ships. But there are individual spooks too. One of the more durable was a wraith that haunted the Pacific for two centuries—and, who knows, may still be haunting that ocean. Identified only by the odd name Ladylips, this gauzy apparition was believed to be the ghost of the captain of a French war vessel, *Ville de Paris*. She was captured by British naval forces during an engagement, then sank after a severe storm-battering while being taken to England as a prize of war.

Ladylips is an incongruous name for this spook, since it usually was reported as having the lower jaw missing. Making its presence more awesome, the specter was characterized by ghastly white skin and an effluvium of dead fish. Perhaps in keeping with the circumstances in which the late French captain's own vessel was lost, his ghost reportedly appeared only during especially bad storms. It's said that over the years sightings of Ladylips in various parts of the

Pacific have been recorded in the logs of a number of American and British naval vessels. He—or it—apparently was seen by many sailors among those ships. A seaman on a U.S. destroyer claimed he spotted Ladylips at the height of a storm off the coast of the state of Washington. His account was published in a magazine in 1931.

Conversion, Human to Ghost

According to the Ladylips legend, the change came about in this way: When the *Ville de Paris* started to go down, the English crew manning her as a prize of war took to lifeboats. In one was the French ship's captain, who had not yet attained ghost status. With their improvised sail the little group had hopes of reaching port or being sighted by a passing ship. To their misfortune, neither hope was fulfilled for a long time. When their food supplies ran out, they tried desperately to catch the sharks circling their boat; and during a struggle with one of the monsters the hapless French skipper met with an accident that tore his jaw off. In unspeakable agony and realizing he was dying, he cut his wrists, ostensibly to provide his own blood as emergency liquid nourishment for his thirsting, starving companions. There was gallantry even between enemies in those days.

Five of the men in the ship's boat survived to eventually reach a remote Pacific isle in the Duke of Gloucester group. And there, in 1928, a crewman from a wandering freighter came across the *Ville de Paris* logbook, in which entries had been made by someone in the lifeboat. From those entries emerged the story of the French captain's conversion. The log mentioned that he was eaten at sea. It didn't specify whether consumption was by sharks or his companions, but a strong inference is that when he slashed his wrists he provided more substantial rations than blood.

A Phantom Photographed

On December 31, 1955, an edition of the Baltimore *Sun* carried a fascinating article. Illustrating it was a startling photograph. The following was the gist of the story as I first heard it:

The ghost of a centuries-dead mariner had been seen, apparently by a few people, aboard the U.S. frigate *Constellation*, a warship of American Revolution vintage that has been restored and enshrined as part of the national monument at Fort McHenry*, Baltimore.

*Its bombardment by the British during the War of 1812 inspired Francis Scott Key to write "The Star Spangled Banner."

One of those who had beheld the specter was Lt. Cmdr. Allen R. Brougham, USN. The phantom was described as wearing a long-outdated ensemble of the type prescribed for officers in the U.S. Navy's earliest years: a rather long, narrow "fore-and-aft" hat, large epaulets, and a sword. Judging by this uniform, the wraith held the rank of captain, or higher. In keeping with apparitions, the ghostly figure was transparent, its appearance made even more eerie by being outlined in a pale blue-white glow. Lt. Cmdr. Brougham was intrigued. He set out to try to photograph the spook.

For counsel on how to best go about this unusual bit of photography he consulted a friend who was an amateur researcher in supernatural phenomena. The friend advised him that an optimum time to catch a ghost with a lens might be at midnight between Christmas and New Year's Day. (Why are spooks invariably most active around midnight? Don't any of them work the day shift?) Accordingly, Lt. Cmdr. Brougham set up a camera overlooking the *Constellation*'s quarterdeck, a likely place where a ship's phantom skipper would appear, and waited.

Within a tick or two of 2400 hours, midnight, the ghost-hunter thought he detected a whiff of gun smoke and heard muffled footfalls moving briskly in his direction. Practically on the dot of 12:00, Captain Phantom strode onto the quarterdeck, enveloped in a bluish-white, phosphorescent-like aura glowing in the darkness. Lt. Cmdr. Brougham clicked his shutter.

The *Sun* reproduced the results. If you will look at the picture, eleswhere in these pages, you will see that it portrays a ghostly figure of a man walking the frigate's deck. The image is transparent and has a blurred quality, as that of all spooks in good standing should have, yet is clear enough for a viewer to make out some of the details of a naval officer's uniform of a distant era.

It subsequently turned out that Captain Phantom was striding toward another revelation.

For reproduction of the picture in this book it was necessary to contact Lt. Cmdr. Brougham. But 22 years had elapsed since the photograph appeared in the *Sun*, and his whereabouts were unknown at the moment. Fortunately, Betsy Nordstrom, the book's editor at the Berkley Publishing Corporation, came up with an address from the Baltimore telephone directory. A letter addressed to Lt. Cmdr. Brougham was answered promptly—and very kindly, I should add—by his son, Allen R., Jr. His father passed away in 1961, he wrote. Then came a denouement.

To begin with, Allen explained, photographing the "ghost" was accomplished by the simple means of a deliberate double exposure. Furthermore, it was taken by day, then doctored to simulate a glowing effect in darkness. And he knows that his father was not the cameraman, Allen said, because he didn't own suitable equipment. He added that the *Sun* article and photograph were intended purely as publicity for U.S. Navy recruiting and for the *Constellation*, which had been brought to Baltimore from Boston four months earlier. The figure in the picture remains unidentified, but has lost its—his—membership in the society of apparitions.

It seems that the *Sun* story and photograph were taken seriously in some quarters. Although not a true believer in ghosts, I was willing to go along myself. It's still an interesting story, but it would have been a *great* one. Despite its denouement, I fancy the idea of a phantom mariner prowling the aged frigate's decks. And perhaps one does, even if it—he—hasn't posed for pictures.

A Ghost Walks the Fo'c'sle

If you can find any very old sailors on the Cornish coast of England you may hear a handed-down story of a seafaring man named Yorkshire Jack and the female specter that haunted him until the day he died by his own hand.

Some century and a half ago in Cornwall there resided a vivacious young lady named Sarah Polgrain. Sarah wasn't exactly a lady, it turned out. A murderess is what she ultimately became. Having wearied of life with a husband many years her senior, she found a new love interest in a chap nearer her age, a sailor called Yorkshire Jack. There was a slight obstacle in this path of roses in the person of Mr. Polgrain. But Sarah found him no problem, really. She removed him as an obstacle completely and permanently by feeding him enough arsenic to kill a team of horses.

Criminology was primitive in that era, but neighbors couldn't help but notice that Mr. Polgrain was conspicuously absent. Moreover, his sudden departure was too convenient to be a coincidence, to their way of thinking. Increasing gossip prompted an investigation, and the body of the late Mr. Polgrain was exhumed for autopsy. The murder plot came to light, and the merry widow was tried and sentenced to be hanged by the neck until dead. In keeping with the times, her execution would be a public affair, open to anyone who fancied witnessing such events.

Granting the condemned woman's last request, her lover was allowed to accompany her to the gallows. Just before the hangman placed the noose around her fair neck, Sarah and Yorkshire Jack entwined themselves in their final embrace in the world of the living. Later it was reported by a witness that her very last utterance to the sailor was a question, "You will?" To which he was said to have replied with obvious reluctance, "I will, I will." The rope cut short any further exchanges.

Sarah Stayed

Making Sarah join the late Mr. Polgrain in death didn't end the matter. At least two people reported seeing her ghost in and outside the village. And Yorkshire Jack saw it quite frequently—too frequently. Instead of bringing him any comfort, however, these visitations caused a marked physical and personality deterioration in the sailor. From a healthy, happy-go-lucky fellow he was transformed into a prematurely aging, sick-looking neurotic, given to looking over his shoulder, muttering to himself, and visiting the local pub with greater frequency. Once, while in his cups, he complained to listeners that "she" was with him always, everywhere he looked. He didn't identify the "she," but there was no doubt in his drinking companions' minds that he was referring to the ghost of the late Sarah.

Then She Left

Yorkshire Jack departed from the hamlet in Cornwall and went to sea. So did spooky Sarah.

Jack's shipmates couldn't help but notice his strange mannerisms. By day he tended to his duties well enough, but he always appeared jumpy and often glanced over his shoulder nervously at something that wasn't there, or at least wasn't visible to others in the crew. By night in the foc's'le his shipmates vaguely but apprehensively sensed a strange, invisible presence in their midst. And all this went on day after day, night after night, throughout a long voyage. Then their ship returned to her home port.

Racked by jitters and beside himself with fear, Yorkshire Jack finally had to confide in someone. He told shipmates about how the ghost of Sarah Polgrain was with him everywhere, every moment, including some nightly room service in the fo'c'sle. With this confidence came a confession of the significance of that brief final exchange at the gallows. Sarah had extracted from him a solemn promise that he would marry her after death, so that they would be bound together forevermore in spirit if not in the flesh—an arrangement of dubious value to Jack, since merely viewing a spook was driving him out of his skull. Yet he dared not go back on his promise. The reason for his sudden confession became apparent when he announced that the date for the wedding was that very night. He didn't explain the reason for the long delay in the nuptials. Maybe Sarah's ghost thought a long engagement more proper.

147

That night in the crew's quarters aboard ship there was no sleep for Yorkshire Jack, and precious little more for his mates, as it developed. With the approach of midnight the tap-tap-tapping of a woman's heels was heard by all in the fo'c'sle. Closer it came, then halted beside Jack. Ashen-faced and trembling with fright, the haunted sailor abruptly left the fo'c'sle and hurried topside to the deck. Right behind him followed the tapping of invisible heels.

On deck, Yorkshire Jack strode straight to the ship's rail and threw himself overboard. The few of his shipmates with the courage to investigate rushed to the rail and peered down into the black water. For a moment or two, they stated later, two ghostly white faces appeared in the dark sea, then vanished. Soon after that, they said, they could have sworn they heard a chiming of church bells, as though signaling a wedding, way off in the distance.

For decades after the haunted sailor's suicide, old-timers in that Cornwall village spoke of hearing distant church bells, chimes for the wedding of Sarah Polgrain and Yorkshire Jack. Some claimed they also heard a ghostly voice promise, "I will, I will!"

Here's Looking at You!

Numerous phantom seafarers have been seen in various parts of the world, and by credible viewers. As you can imagine, though, it's an extremely rare occasion when a beholder can support his claim with a photograph.

Such an occasion was first described in *Service*, company house organ of the Cities Service petroleum people in February of 1934. Later accounts appeared in a book and magazine. It was introduced to me by Vincent Gaddis's absorbing volume, *Invisible Horizons* (Chilton Book Company, Radnor, Pennsylvania, 1965), after which I wrote to Cities Service headquarters in Tulsa, Oklahoma. Perhaps true psychics and experts in supernatural matters can explain it. It's beyond the ken of laymen.

Late in 1924 the Cities Service tanker *Watertown* sailed from San Pedro, the harbor of Los Angeles, en route to the Panama Canal and then New Orleans. In a tragic accident two members of her crew, Michael Meehan and James Courtney, were asphyxiated by gasoline fumes; and on December 4th their bodies were committed to the deep off the coast of northern Mexico, at a point where the Pacific was nearly 1,500 feet deep. Interment occurred at sunset, a time which may have had some psychic significance in light of what followed on the next day.

On December 5th, just before dusk, the tanker's first officer was startled to see the ghostly faces of the two deceased seamen amid the waves off the vessel's port rail, just about opposite where their bodies had been released to the sea. Somewhat enlarged and seeming to float like flat portraits, the apparitions were about ten feet apart and roughly forty to fifty feet beyond the ship. The dead men's features were readily recognizable and therefore identifiable as the visions kept pace with the *Watertown*.

Thereafter the heads appeared daily, between late afternoon and dusk, always in the same general location with respect to the tanker. They would be visible for approximately ten seconds at the most, then fade, then reappear. Before the *Watertown* reached Panama nearly all her men had seen the phantom faces of Mike Meehan and Jim Courtney. It occurred to the first officer to try to get a picture, but unfortunately no one aboard had a camera. When the vessel left the Pacific Ocean to enter the Panama Canal, the apparitions ceased to appear.

This Time, a Camera

In New Orleans the *Watertown*'s master, Captain Keith Tracy, and her engineer, Monroe Atkins, reported the eerie goings-on to Cities Service's regional office in that port. There an executive of the corporation was especially intrigued by their report and learned with disappointment that no attempts had been made to photograph the supernatural phenomenon. The skipper was determined to try to correct that situation. On the outside chance that the apparitions might appear again when the *Watertown* re-entered the Pacific on her return to California the first mate bought a camera and had it checked to make sure it was in perfect condition.

Sure enough, when the ship was back in the Pacific the ghostly heads appeared as before. Captain Tracy personally shot six pictures and locked the exposed film in his safe. Later the company executive in New Orleans who was so interested in the affair took this film to New York, where it was processed professionally. Five clicks of the camera shutter had drawn blanks. The sixth exposure jolted its viewers.

It showed two ghostly faces: Not sharply defined, understandably, but discernible and unmistakable. The one on the left was identifiable as Courtney's. A few feet away and a bit clearer, the face of bald-headed Mike Meehan looked up at the lens. Prints and negatives were examined minutely by photography technicians of the famous Burns Detective Agency, who reported they could find no evidence whatsoever of fakery. The astonishing picture subsequently was exhibited in the lobby of the Manhattan, New York headquarters of Cities Service.

149

There are records that show that the faces of Mike Meehan and Jim Courtney were observed in the sea during two or three Pacific runs of the *Watertown* after their demise, but they appeared with decreasing frequency. There doesn't seem to by any record of their being seen thereafter.

To term the apparitions some kind of supernatural illusion is a vague "explanation" at best, and really doesn't explain them at all. But then, by human reckoning there is no ready explanation. All we can say with certainty is what the apparitions were *not*. They were not merely optical phenomena, since they appeared too often, were too constant in details and position, and always portrayed identifiable features of the deceased seamen. Nor were they hallucinations. Proof was the photograph. Even without it, the ghostly faces were seen by too many witnesses, individually and in groups.

In addition to the cause, we can ponder these queries: Why, do you suppose, were the phantom faces seen only in the sea, and always in approximately the same place, relative to the *Watertown*? Why not aboard their ship too, instead of in the water? And why were they observed only in the Pacific Ocean, where the seamen had been interred, and not in other waters, such as the Gulf of Mexico, transited by the ship?

20

Mystery of the Dry "Wet"

One of my favorite sea mysteries (a bit of stretching is required to fit it into that category, but no matter) is a little saga that took place on the expanse of a large bay on the southern shore of Long Island. Its hero was never identified, but the incident never ceases to amuse me.

On a summer morning an expensive automobile tooled into the parking area of a fishing station and rowboat livery beside the bay. Out stepped its sole occupant, a man of perhaps middle age. It was rather late in the morning for a customer to rent a rowboat to go fishing. Most livery patrons are on the dock bright and early. Yet a late arrival didn't surprise the station's proprietor, a fellow I'll call Don. He was used to patrons dropping in at all hours: anglers who like to go out at their convenience or sandwich in a little fishing time whenever they can. Besides, a customer is a customer, and who's going to subject him to a psychological quiz?

What did give Don pause for thought, however, was this customer's attire. He was dressed to the nines: Carefully creased white trousers, natty blazer, and a yachting cap. This ensemble was too fancy for an establishment where T-shirts, nondescript pants, beat-up hats and six-packs of beer were standard. Don's curiosity deepened as the man approached in a dignified stride. He carried no fishing tackle. In fact, he was completely empty-handed. This *was* strange, since the station's clients always arrived toting fishing tackle, miscellaneous gear, and lunches. "Maybe this guy is gonna buy some tackle," Don said to himself hopefully.

But the stranger didn't. He didn't buy anything. He did, however, rent a rowboat and motor for the day. Don noted that he seemed to be a man of means and breeding, but curbed his inquisitiveness as he took the fellow's money. When last seen, the stranger was sedately putt-putting toward the open bay that sparkled in the morning sun for several miles to the west, south and east.

In his daily chores around the place Don quickly forgot the dude.

Weather and Worry

Along about midafternoon a squall whistled in over the bay. Rain fell in torrents, lightning flashed and claps of thunder assaulted the

151

ear. The wind blew in gusts to forty knots. Even ducks could have been in trouble. Many of the fishing station's customers saw the squall coming and beat it to the dock. For others it was a dead heat. Still others came in drenched to the skin.

The little storm lasted no more than about half an hour, but it was brisk. When it had blown itself elsewhere Don anxiously checked his rowboats. Fishing station operators feel a responsibility for the safety of their customers. All Don's boats were in . . . no, wait, all except one. You guessed it, the missing rowboat was the dude's.

Maybe he beached the boat to wait out the storm, Don told himself, or he could've had motor trouble. A couple of grim possibilities also entered his head, but he'd give the stranger just a little while longer. Don glanced southward toward the bay. Not a boat in sight. He'll probably be along any minute now, the station operator tried to convince himself.

But when a reasonable margin of time passed and the missing rowboat still hadn't appeared, Don became very concerned. Now that he thought about it, the guy really didn't impress him as being a boatman, yachting cap notwithstanding. Besides, people have been known to ignore or underestimate weather's danger signals on the water. Don decided he'd call the Coast Guard and join in a search with his own skiff. He was just about to put this rescue plan into action when his ears caught the unmistakable sound of an outboard motor still some distance off. Don hung up the telephone and rushed to his dock.

Here came the missing rowboat. Don could recognize his red and white hulls from way off. His shoulders slumped in relief.

He fully expected to see the stranger looking like a drowned rat and the rowboat half-filled with water. When the boat came alongside the station's float, Don's relief gave way to astonishment. Both dude and boat were bone-dry. His entire outfit, yachting cap to white shoes—even the creased trousers—were as immaculate as when he left hours earlier. The stranger was dry outside, but inside was something else. Fact was, he was stoned out of his skull. Speechless, Don watched him tie up the rowboat deftly, flash a silly grin, touch the peak of his cap in salute, then stride with unsteady dignity to his expensive car. By the time it occurred to Don that this guy was in no condition to walk, let alone drive, he was gone.

No Answers

Don never saw him again. Nor does he have the foggiest notion who he was. But every once in a while he still grapples with these questions:

1. How did the guy stay completely dry, out on open water where

152

there was no shelter of any kind, during a torrential downpour? And in a strong wind, too.

2. Even more interesting, how did he get drunk? He was sober when he left in the morning, and took nothing with him. And in the area where he had been there are about as many taverns and liquor stores as there are, say, giraffes.

I've toyed with the theory that somewhere out on the bay our hero met friends with a cruiser, whereupon he whiled away the squall in comfort with numerous libations. The flaw is that it doesn't explain the dry rowboat.

You might suggest that he beached the rowboat, managed to work her onto the sand, then flip her over and crawl underneath for shelter. That's possible, I suppose, if he had enough strength. The flaw is that it doesn't explain how he got stoned.

It appears that the dry "wet" will remain forever one of the sea's unsolved mysteries.

21

The Pacific's "Mary Celeste"

On a November morning in 1955 a British interisland ship called
Tuvalu was plodding along nearly 500 miles southwest of the
insular Territory of Western Samoa when she came upon a badly
listing motor vessel, wallowing dead among lazy swells and
troughs. Obviously a derelict, she was an unidentified drifter at
first. Although she still rode fairly high in the water for her condition, she canted so badly to port that no name could be detected on
that side. But there was no problem reading the letters on her
starboard bow. Large and black, they spelled *Joyita*. It was a name
that soon would be heard around the world.

Accounts disagree as to whether or not the *Tuvalu*'s master,
Captain Gerald Douglas, dispatched a boarding party to the derelict, and that detail remains unclear. No matter. In either event, a
superficial inspection established these facts: (1) There was no one
aboard the *Joyita*, at least not in evidence above deck. However, the
possibility of corpses in the wandering vessel's submerged cabins
was not discounted. (2) In addition to being abnormally low in the
sea and listing badly enough for her port rail to be under water, the
Joyita had suffered appreciable damage topside, such as could have
been inflicted by very strong winds and/or heavy wave-battering.
Part of her superstructure—that is, structure above deck—was
missing. (3) Abaft her bridge—an open, railed affair atop her
forward cabin—a canvas tarpaulin appeared to have been jury-
rigged as an improvised substitute for a missing portion of her
damaged superstructure. It could have been an awning against the
sun, but more likely was for catching rain for drinking water. This
was interpreted to mean that someone was aboard after the *Joyita*
encountered whatever it was that put her in this fix.

The vessel's logbook, which could have supplied valuable clues,
was never found.

At least the *Tuvalu*'s discovery answered a month-old question
about the *Joyita*'s whereabouts, something which an extensive
search had failed to reveal. But the discovery also precipitated
several queries which could not be answered.

Word of the finding was radioed to Suva, capital of the Fiji

Islands, on Viti Levu. A tug was set to retrieve the *Joyita*, and towed her some ninety to a hundred miles to a harbor at Vanua Levu, second largest of the Fijis and thirty-eight miles northeast of Viti Levu. There water was pumped out of her hull. No bodies, bloated and fish-belly white, were in the previously submerged cabins. Deepening the mystery, the hull had neither holes nor signs of an explosion or collision. In fact, it looked quite sound structurally.

From Vanua Levu the *Joyita* was towed to Suva, where she was hauled out of the water for a thorough physical examination. During one of the inspections it was learned why the derelict had remained afloat: Considerable cork filled hull spaces, making her virtually unsinkable.

A second search was launched, this time for bodies and possible survivors. It drew a complete blank. Meanwhile, authorities set about determining who had been aboard and gathering information for an official inquiry. The story of finding the *Joyita*, dead and deserted, became news all over the world. Wire services flashed it everywhere. Millions of readers, the author included, were fascinated by newspaper accounts and the spread *Life* magazine (December 12, 1955) devoted to it.

A Kaleidoscopic Career

The *Joyita* measured sixty-nine feet, stem to transom, and had a gross weight of seventy tons. Diesel engines turned her twin propellers. Although she still retained traces of her trim lines, her forlorn appearance on discovery belied her beauty when young. She was born in 1931 as a private yacht in a California boat builder's yard. Strong wooden-hull construction embodied graceful good looks; smart appointments characterized her interior. Her christened name summed her up. *Joyita* is the diminutive of *joya*, Spanish for "jewel," and means "little jewel." Under the burgee of her first owner, a former motion picture director, she graced U.S. West Coast waters through her early years, often, it was said, with a galaxy of Hollywood's brightest stars aboard.

Then, in 1936, began a procession of owners, and *Joyita*'s years as glamor girl were numbered. Her second owner enjoyed her as a pleasure craft until World War II, at which time the U.S. Navy enlisted her as a patrol boat. After the war she was acquired by a firm in Hawaii and converted to a commercial fishing vessel. Installed were twin 225-horsepower diesels, refrigerated holds, and that cork flotation mentioned earlier. In 1952 came yet another change in ownership. Why she purchased the *Joyita* isn't clear, but the next possessor was a woman educator at the University of

155

Hawaii in Honolulu. A logical reason for the purchase would seem to be a business investment, for this lady rented the motor vessel to a friend, Captain Thomas H. Miller. He, in turn, put the former yacht back into commercial fishing service. Registered in the port of Honolulu, *Joyita* flew the American flag, although Captain Miller was a British subject.

Captain Thomas H. Miller was to become the leading character in the mystery I call "The Pacific's 'Mary Celeste.'" Whatever happened aboard the *Joyita*, he was one of twenty-five people who vanished from her without a clue.

Troubles Begin

Captain Miller has been described as an affable Welshman, a colorful fellow but a competent skipper. Having chartered the *Joyita*, he launched a commercial fishing enterprise in Hawaii. Perhaps his talent as a commercial fisherman was not commensurate with his ability as a captain. Or he could have experienced an unrelenting series of bad breaks. The oceans do not teem with fish to the degree that most people believe; and the fortunes of trawlers can go up and down like the tide.

Captain Miller seems to have had modest success at first, but then it was all uphill, complicated by malfunctioning of the *Joyita*'s refrigeration equipment which caused spoilage of catches. When he shifted operations to Pago Pago, capital of American Samoa, his luck didn't change for the better. The unfortunate skipper sank deeper and deeper into debt until he was dead broke. Authorities in Pago Pago finally seized certain of the *Joyita*'s documents, apparently in an effort to assure payment of mounting debts while preventing the skipper from selling a vessel that wasn't his.

In March 1955, Captain Miller took the *Joyita* from American Samoa to Apia in Western Samoa, which was under the New Zealand trusteeship. There he entertained hopes of chartering the *Joyita* for one purpose or another to realize an income again. A possibility was a steady charter to the Western Samoa government as an interisland supply boat and ferry. His luck continued to be consistently bad. He was refused a government charter because he lacked those ship's papers still held by American authorities in Pago Pago. And, being broke, he couldn't finance commercial fishing operations.

For months the *Joyita* lay idle at Apia, while things went from bad to worse for Captain Miller. He lived on board alone, his crew having become discouraged by a noticeable lack of pay checks. His existence was hand-to-mouth, with survival only by help from friends and what little money he could earn from occasional part-time work on shore.

Tom Miller's luck had reached its nadir when there came an unexpected break in September of 1955. Pressured by a growing need for food staples and medical supplies in the Tokelau Islands, part of the Western Samoa trusteeship, and by need for transportation of copra*, a major Tokelau export, a company in the copra business arranged to charter the *Joyita*. This windfall came about through the efforts of a New Zealander named R. D. Pearless, newly appointed district governor of the Tokelaus. He and Captain Miller had become friends in Apia. Ironically, he was to be one of the persons mysteriously vanished from the *Joyita*.

What with his personal finances and long inactivity, it wasn't easy for the skipper to enlist a suitable crew, even with a charter nailed down; but he managed. "Chuck" Simpson, an experienced seaman of American Indian extraction who had settled in Western Samoa, signed on as mate. Chuck fulfilled the requirement that a commercial vessel of American registry have an officer who is a U.S. citizen (Captain Miller was a British subject, remember). Also enlisted were two former crewmen, Gilbert Islanders named Tanini and Takoka. The former was hired as the *Joyita*'s engineer, the latter as her bos'un. Both men reportedly were devoted to Captain Miller.

After months of idleness the *Joyita* wasn't in the best of health. Her crew worked like beavers to get her into shape, especially her propulsion equipment. It came out later that Captain Miller had experienced some kind of trouble with the portside diesel's clutch. But at the time of the Tokelau charter he didn't consider this too serious and figured it could be repaired on the way out. When the *Joyita* was discovered as a derelict, the port engine's clutch was partially disconnected, indicating that the vessel had put to sea on one engine.

Captain Miller seems to have been conscientious and thorough in most matters, so it's incongruous that he should be slipshod in anything so vital as his vessel's radio. Yet evidence points to just that. On previous sailings in the islands *Joyita* was lax in maintaining radio contact with Apia, a fact which now was brought to the skipper's attention. She was assigned the call letters WNIM, and an authority in the islands' radio setup urged Miller to test her transmit-

---

*Copra is the dried meat of coconuts. For years its oil has been used in the manufacture of vegetable margarine, cooking fat, soap, shampoo and other products.

ter before departure, then keep in touch with Apia daily between 1000 and 1600 hours, 10:00 A.M. and 4:00 P.M. The skipper seems to have ignored both suggestions. Once she departed on what was to be that tragic voyage, Apia never heard from her.

Aloha

The *Joyita*'s destination was the port of Fakaofo in the outlying Tokelau Islands to the north of Apia. It wasn't a long trip as voyages go, only about 270 miles. The estimated time required was forty hours, give or take. Stowed aboard at Apia were medical supplies and a heterogeneous payload of other items needed in the islands.

In addition to Captain Miller's friend R. D. Pearless, passengers included a physician named Parsons, and J. Hodgkinson, an employee of the hospital in Apia (both accompanying the medical supplies to supervise their distribution in the Tokelaus); a G. K. Williams, retired executive of a New Zealand insurance firm; and his companion, J. Wallwork. Williams and Wallwork were aboard as representatives of the copra company chartering *Joyita*. Williams reportedly carried a sizable sum of money with which to buy copra in the islands. This would be the cargo on the return run. All the others passengers were native islanders. Two children were among them.

The *Joyita* departed under some clouds:

1. There was doubt, in certain minds at least, about her engines' reliability.

2. Some could question the reliability of her radio.

3. Under terms of her registry, the *Joyita* was not supposed to carry passengers for pay. Miller had circumvented this ban in the case of R. D. Pearless by listing him as supercargo (ordinarily, a ship's officer who represents her owners and has charge of the cargo and business affairs during a voyage). How he explained all the others—if he did—isn't clear.

The *Joyita* first set out from Apia on the morning of October 2nd. Watching from his home overlooking the harbor was an employee of the local government's marine affairs department. As he looked on he saw the vessel suddenly belch black smoke from her exhausts, then drift helplessly until an anchor was dropped. Apprehensive, he immediately telephoned Acting High Commissioner T. R. Smith to report the incident and suggest the advisability of taking off the government representatives. Along with Pearless, Dr. Parsons and Hodgkinson were considered such. Since the *Joyita* was at anchor, they decided to wait.

Engine trouble continued to plague the vessel. Expecting to sail that evening, some passengers went ashore and found a club where they could wait out repairs in more pleasant surroundings with liquid refreshment. It wasn't until early the next morning that *Joyita* got under way again. By then her crew was fatigued from working on the engines all night, and the passengers were in a sour mood. The government representatives had not been removed.

Sometime around 5:00 A.M. the *Joyita* slipped away from Apia. Her skipper probably sensed that he would have been stopped if he had still been in the harbor later that morning. At the official inquiry after the tragedy, Acting High Commissioner Smith explained that he okayed the trip to Fakaofo as a mercy mission to transport urgently needed medical supplies in the Tokelaus and to provide a physician to survey the islands' health conditions.

There apparently was never any question about Captain Miller's ability. However, a sidelines observer might wonder if his desperate need for income may have driven him beyond common sense. After all, this was a chance for success after many setbacks. Too, he did have firm faith in his vessel's unsinkability, due to her cork flotation—justified by the fact that she was still afloat when discovered; and the trip was not an especially long one.

Pieces of the Puzzle

The *Joyita*'s fateful voyage began on October 3rd, five weeks before her discovery as a derelict. Based on an estimated amount of fuel already consumed, it was believed that whatever befell her happened on that first night out. Her electrically powered clocks had stopped at 2253 hours, 10:53 P.M. The switches of her riding and navigation lights were in the "On" position, indicating that she had been moving in darkness. Upwards of 2,500 gallons of fuel sloshed in her tanks, sufficient to give *Joyita* a cruising range in the vicinity of 3,000 miles. She carried more than enough food and water for the trip. Her radio gear was tuned to 2182 kilocycles, an emergency frequency; but a later shoreside examination revealed a break in the lead to its transmitter antenna, which would have prevented her signals from being heard any farther than a mile or two. There was no way of determining whether this break occurred before, during or after the craft's trouble.

Missing were three lifesaving flotation craft, life jackets, compasses and sextant, vital to plotting courses in the vast Pacific, and, as mentioned earlier, the vessel's log.

These and more details that came out during the official inquiry in Apia in February the following year only deepened the *Joyita* mystery.

Since an extensive sea-air search conducted after the derelict's discovery failed to find any survivors or bodies, there was no clue to the cause of their vanishing. Also strange was the disappearance of her cargo without a trace. On deck she had carried about 2,000 board feet of lumber and several large but empty oil drums. In an emergency some of the lumber, and perhaps a few of the drums, could have been fashioned into a crude raft; but it wouldn't account for all that material. Further, whether the lumber had been jettisoned to lighten the vessel or had broken loose and been swept overboard by a rough sea, there should have been some around as flotsam; yet none was ever found.

Even harder to explain is the reported disappearance of cargo from her holds. It came out that *Joyita* was carrying approximately seventy sacks of food staples—rice, sugar, flour and the like—for the Tokelau Islanders. Weighing in the range of fifty to 150 pounds apiece, they were missing from the after hold. In another hold when she sailed were cases of aluminum strips, used to protect coconut trees from rats. They were gone. Although accounts do not mention it being given serious consideration, one wonders if these items were tossed overboard to lighten the vessel in an emergency.

What Did Happen?

Investigators attacked this question from every conceivable angle. Theories sprang from both inside and outside the circle of official inquiry.

Most chilling of the outside speculations was that a "pirate submarine" of unspecified nationality attacked the *Joyita*, killed or captured everyone aboard, then looted her of cargo. This theory was inspired by reports of a strange submarine, or submarines, in the general region of the Fiji Islands. The reports were unofficial, and some observers wrote them off as cases of mistaken identity, the "submarines" actually being whales lolling at the surface.

The piratic submarine suggestion received other impetus when the sixty-foot motor vessel *Arakarimoa* vanished mysteriously with her crew and passengers during the night of December 28, 1955, while en route from Tarawa to Maiana in the Gilbert Islands. This became a kind of side-mystery, you might say, in the *Joyita* case. The missing vessel was described as being in good shape all around. Moreover, when she vanished she was in company of a sister vessel, *Aratoba*. The two were in sight of each other at least until midnight of the 28th, after which the *Aratoba* inexplicably increased her speed and passed from the other vessel's sight, reaching port safely the next morning. Nothing is mentioned about any radio communications between the two during those midnight hours. A sea search instituted on December 30th failed to turn up any trace of the *Arakarimoa* and her people.

In Fiji a major newspaper hit the streets with a headline charge that everyone aboard the *Joyita* had been murdered. The accompanying story suggested that the vessel had innocently come upon a Japanese fishing fleet known to have been working in an area along the *Joyita*'s course to the Tokelau Islands. The speculation went on to say that perhaps the stranger had seen something the Japanese fishing fleet didn't want her to see, and that she had been stopped, boarded, and her passengers and crew either taken prisoner or killed. Although Japanese commercial fishing in their waters was resented strongly by the Fijians, officials of the islands' government were quick to shoot down the newspaper's theory as pure fantasy.

A plausible but unverified theory was that the *Joyita*'s people were victims of a seaquake—an undersea earthquake—a natural phenomenon in the Pacific and often accompanied by catastrophic waves. Earlier that year a vessel not far off Fiji was struck by such a submarine disturbance, so sudden and violent it tossed crew and passengers into the sea. However, a check of other ships in the *Joyita*'s area at the time of her disappearance brought no report of a seaquake. Of course, there remained a distinct possibility of a freakish, localized high-wave pattern striking the *Joyita* and spilling her occupants into the ocean. It seemed unlikely, though, that all twenty-five souls would have been so jettisoned at once. Pursuing that theory anyway, if the calamity occurred while the *Joyita* was under way, odds would have been high against anyone being able to get back on board.

It also was suggested that she had been smacked—or sideswiped —by a waterspout, a phenomenon at sea that could be likened to a cyclonic column of water moving in the manner of a tornado. In support of this conjecture it was pointed out that the *Joyita* lost part of her superstructure on the port side and bore other signs of mauling. Seasoned mariners agreed that a waterspout could have damaged the vessel, but in this instance it was not enough to cause everyone to abandon her. A countertheory argued that incessant battering by waves could have caused the damage if she were abnormally low in the water and listing to port. That she had been listing for some time was indicated by barnacles already collecting high on her port side.

Mutiny sounds more far-out than piracy. There was no evidence to support it. On the contrary, Tokoka and Tanini were supposedly devoted to Captain Miller. Besides, if they had wanted to quit, why wait until the *Joyita* was at sea? They had a perfect opportunity when she was still in Apia harbor with engine trouble.

And there persisted that fact that the vessel was still afloat, without signs of fire or explosion.

The *Joyita* became a *cause celebre* with publicity, unofficial observers and amateur investigators all over the world.

Detailed examination on land lifted a mantle of mystery from the cause of her condition when found at sea, or at least it offered the most sensible explanation. It was discovered that a section of pipe in the portside engine's cooling system (the same diesel with the clutch trouble) had deteriorated so badly that it allowed sea water to pour into the vessel. Although the pipe was only about one inch in diameter, over a period of time the leak could have been enough to cause serious flooding. To make the defect even more insidious, the leak was hidden underneath the engine compartment floor, and the diesels' drone would have masked the sound of inrushing water.

The crew had to become aware of a bad leak, but the sea's invasion may have already passed a point of no return by the time it was detected. A few mattresses in the engine compartment bespoke emergency efforts to stem the flood, but it appeared that no one was able to pinpoint its source. Meanwhile, the *Joyita*'s pumps were unable to cope with the situation. Later inspection showed their efficiency to be seriously impaired by accumulation of gunk in their suction pipes. The crew had jury-rigged—improvised, that is—an auxiliary pump in a last-ditch attempt to assist the regular pumps, but either this proved inadequate or was put out of commission before it could do any good.

Before long, water rose high enough to drown the engines and electrical system. Now the stricken *Joyita* was helplessly adrift and in darkness, unable to radio for help.

After hearing the testimony of marine experts who examined the derelict on shore, the panel of inquiry officially declared the cause of her predicament as flooding by a defective pipe in the port engine's cooling system.

The Biggest Question Still Unanswered

Investigators had come up with a sound, plausible reason for *Joyita*'s condition when found. But no one had so much as a shred of substantial evidence concerning the fate of her passengers and crew. That is, there was nothing conclusive as to *why*, or *how*, they left the vessel. Failure to find any bodies in the sea is not particularly mysterious. The Pacific Ocean is an incredibly vast expanse, as you know if you've ever flown across it. The Pacific also is well populated by sharks, to which the author can attest by having been in Tahiti, the Fiji Islands, and Australia.

One of the case's several blind alleys was why Captain Miller

162

would abandon *Joyita*. It seemed strange, since he had outspoken confidence in her unsinkability. Further, he knew, as modern pleasure boatmen with flotation-filled craft are advised, that it's generally safer to stay with a vessel as long as she remains afloat, even if her deck is awash, than to strike out on one's own. (That is, barring fire or explosion, of course.) That policy is doubly sound in shark-infested waters. For the same reasons, the crew would have stayed with the *Joyita*, and Captain Miller undoubtedly would have advised his passengers to do likewise.

Another puzzler was why a seasoned mariner such as Captain Miller hadn't fashioned sea anchors. These could have been improvised from his deck cargo of lumber or from tarpaulins and would have helped to stabilize his vessel in rough water. They might also have prevented her portside superstructure from being damaged.

After the official inquiry closed, an interesting and possibly very significant detail came to light. Because the marine expert involved wasn't called as a witness, it didn't appear in the testimony. While going over the *Joyita* on shore, this expert found a stethoscope, scalpel, sutures, and bloodstained bandages among some debris on board. Someone had been injured and was treated by Dr. Parsons, that much was certain. But who? Was it Captain Miller? If so, that might explain a couple of things. If the skipper had met with incapacitating injuries, or worse, or had been knocked unconscious, perhaps the passengers, and a crewman or two as well, fled the *Joyita* in panic. That would account for the missing lifesaving devices. The medical instruments and bloodstained bandages hinted strongly at serious injuries, whoever was the victim. That brings up another potentially significant question, especially if the victim happened to be the skipper: How had the injuries been sustained? In a fall? Or perhaps in a fight?

Theorists leaned toward pinpointing Captain Miller as the victim. Badly injured, he might not have been able to abandon ship even if he wanted to. Unconscious, he couldn't. In their haste the others may have left him for dead. You'll remember that the jury-rigged awning indicated someone having been aboard the derelict for a while. All things considered, speculation points to that someone being Captain Miller. And since injury may have prevented him from rigging the tarpaulin, at least one person must have remained aboard with him. It was believed that the person had to be either Tanini or Tokoka. Or maybe both stayed. As for Chuck Simpson, the mate, it seems odd that he would have left, knowing of the *Joyita*'s buoyancy, yet he may have been the one who carried off the vessel's navigation instruments and logbook.

Of necessity, "ifs," "possibles," and "maybes" riddle the *Joyita* case.

If the passengers and mate abandoned ship, with or without the skipper's blessing, they might very well have been swallowed by the Pacific, especially if rough seas were running. If Captain Miller and others remained aboard, their ultimate disposal is impossible to explain in light of what is known.

As this particular speculation saw it, here's what could have happened:

Sometime after her mishap, whatever it was, the now-helpless, drifting *Joyita* was spotted by a passing vessel, type unknown. When the stranger received no reply to hails, she sent a boarding party. Once on the *Joyita*, the idea of looting her was conceived. The invaders had no problem with Captain Miller, since he was helpless or perhaps already dead. Whoever was with him probably resisted the looting, and was promptly eliminated. After dispensing with everyone aboard, the strangers removed everything they could salvage from the derelict, including her navigation instruments— perhaps her log too. If Chuck Simpson had remained aboard, he was among those erased.

But that was only another conjecture. Fact is, we still don't know for sure what actually happened to the unfortunates on the "Little Jewel," and probably never will know.

In the Wake of a Derelict

According to a wire service item, *Joyita* was auctioned off to a Fiji Islands planter in July of 1956. Unless the price were an exceptional bargain, it's a wonder anybody would want her after all the grim publicity she recei ed. Even normally unsuperstitious people had a right to wonder if he poor thing carried a Jonah or jinx. This thought was given further nourishment six months later when she struck a reef and was abandoned again, but not under mysterious circumstances this time. One follow-up says she was then taken to Levuka on Ovalau, one of the Fijis. By now, Fijians looked on the *Joyita* not only as a hoodoo, but also as a haunted ship, tenanted by the ghosts of the twenty-five people who had vanished.

Another official curtain rang down on the *Joyita* affair in 1961. In that year Captain Thomas H. Miller was declared legally dead.

But the saga didn't quite end there. A writer named Robin Maugham became deeply interested in the *Joyita* mystery and journeyed to Fiji for a personal investigation. Far from bothering him, what the Fijians said about the ill-starred vessel intrigued him all the more; and in the early 1960s he purchased her. Joyita's fate from that point on is, simply, unknown.

22

Flotsam & Jetsam

In a preface to his immortal *The Adventures of Huckleberry Finn*, Mark Twain warned: "Persons attempting to find a motive in this narrative will be prosecuted; persons attempting to find a moral in it will be banished; persons attempting to find a plot in it will be shot."

In other words, relax and enjoy it.

Same goes for this section of your book—and the entire book, for that matter.

Here, for changes of pace, is a medley of odd and off-beat fragments of maritime history.

A Shark Turns Detective

All kinds of plots have been spawned at sea: Piracy, mutiny, barratry, and murder. In many instances the last chapter was discovery and retribution. An untold number may have gone undetected and unpunished. And then there are plots with surprise endings. One of the strangest of these occurred in Australia in 1935.

A large tiger shark—a dangerous species in any waters and greatly feared in Australia—was captured alive and placed on exhibit in a Sydney aquarium. About a week later the toothy beast suddenly regurgitated a human arm, a little performance not usually included in the price of admission. Evidently the arm had been stored temporarily in the monster's stomach, where—and this is odd in itself—it had not yet been attacked by the shark's powerful digestive juices.

It was so well preserved, in fact, that medical examiners could make out a tattoo. Now, tiger sharks are as noted for their scavenging as they are for their people-killing, so this arm could have as readily been chomped off a corpse as off a living victim. However, further examination revealed that it had not been bitten off by the tiger, as would have been detectable by the amputation's jagged nature, but had been cleanly severed by a knife.

Accordingly, the matter was turned over to the police. They had very slender leads, but the tattoo helped; and police traced the limb to a boxer who had been reported missing a couple of weeks earlier. Aided by other leads picked up along the way, detectives assembled this picture.

The pugilist had been part of a conspiracy to wreck a yacht for insurance—in maritime law that would be termed barratry. Somewhere along the line their scheme went askew, whereupon the boxer was murdered by his fellow plotters, undoubtedly to prevent his talking. Police figured that they attempted to jam their late companion's remains into a box for disposition at sea, but the arm wouldn't fit, so they cut it off and dropped it overboard separately. Along about then the tiger shark came along and gulped it down.

Thus did a fish collect and preserve evidence leading to a murder trial.

Study in Contrasts

What may remain for all time the worst marine disaster ever involving a single ship—or even two—is the sinking of the World War II German war vessel *Wilhelm Gustloff*. On January 30, 1945, she was torpedoed off Danzig by the Russian submarine S-13. Approximately 7,700 lives were lost.

In dramatic contrast, in a sinking of the largest ship ever to be lost, there was not a single fatality. In December 1969, the 206,600-ton Royal Dutch Shell supertanker *Marpesa* was on her maiden voyage from Rotterdam. Off the shores of Dakar, Senegal, she was torn open by an explosion and went down the next day.

Never Say Die

In the autumn of 1829 a schooner named *Mermaid*, manned by a crew of eighteen under command of a Captain Samuel Nolbrow and carrying three passengers, set sail from Sydney, Australia, for Hong Kong. While negotiating Torres Strait between Australia and New Guinea, a passage notorious for its reefs and treacherous bars, the *Mermaid* lost an argument with a submerged coral mass and sank. Crew and passengers escaped with their lives by clinging to rocks.

A few days later they were rescued by the bark *Swiftsure*, passing through the strait. Two days after that, the *Swiftsure*—swift, perhaps, but not so sure, as it turned out—ran aground and was wrecked. Fourteen members of her crew and the twenty-one from the *Mermaid* survived by swimming ashore.

There they were marooned until a schooner called *Governor Ready* happened to come along. All the castaways were taken aboard, joining the host vessel's thirty-two men, and the schooner headed for New Guinea. While en route a fire erupted on the *Governor Ready* and burned her to the waterline. Now there were sixty-seven survivors from three vessels, eluding death by taking to the schooner's boats.

Another vessel wandered by, the cutter *Comet*. She rescued the survivors of the *Mermaid*, *Swiftsure*, and *Governor Ready*. Not long afterward, the *Comet* was clobbered by a violent storm and foundered. She added her twenty-one men to the others trying to stay alive. Floating around, in whatever boats they had been able to launch, or clinging to wreckage, were eighty-eight people in all. Not one life had been lost so far.

Now enter the *Jupiter*, manned by a crew of thirty-eight. She plucked the eighty-eight survivors from the sea. She had barely begun to adjust to this large increase in her human cargo when—are you ready for this?—she climbed upon a coral reef, stoving a yawning hole in her bottom. Tragedy loomed for 126 souls from the *Mermaid*, *Swiftsure*, *Governor Ready*, *Comet*, and *Jupiter*. But again Lady Luck followed a frown with a smile. All hands escaped death by clinging to rocks jutting out of the sea.

By and by, along came the *City of Leeds*, a schooner. She added the 126 to her consist of 100 people.

At last fate wearied of the game and threw up its hands. The *City of Leeds* fetched port safely.

Animal Sense

You've heard the old saying that rats will desert a sinking ship. We can update it by adding that maybe those loathsome vermin have ESP.

The *Paris C. Brown* was a Mississippi River paddlewheeler plying between Louisiana and Ohio. One July day in 1889, while she was tied up at a wharf in Plaquemine, Louisiana, two large rates were observed scurrying down the vessel's gangway to shore, obviously in haste to get off. This exodus was noted immediately by three superstitious crewmen, who figured that maybe the rodents were giving them a message. They promptly left the steamer hard on the heels of the rats.

The superstitious trio became the only survivors of the *Paris C. Brown*, which not long afterward vanished without a trace.

Whale Tales*

Even before the decline of American whaling, hastened by discovery of petroleum in Oil City, Pennsylvania, circa 1850, whaling ships from such U.S. ports as Sag Harbor, New York, and New Bedford, Massachusetts, had begun to find their business growing tougher and tougher. Such had been the slaughter of the big marine mammals that hunters had to venture farther and farther at sea to find their decreasing numbers.

Such voyages took the American flag to remote waters all over the world—Arctic, South Seas, and Antarctic. For some vessels it was necessary to be away for as long as three or four years before a worthwhile cargo of whale oil was accumulated, and sometimes they didn't collect a really paying cargo even then. One of the longest voyages ever undertaken was that of the New Bedford whaler *Barclay*. She was gone five years and five months. The record doesn't comment on her success or lack of it, but it would seem from the span of time involved that she found slim pickings.

When American whaling was in its halcyon years of the first half of the 1800s it was common practice to sell shares in a voyage. For investors it was a business opportunity, but also a gamble. Money so collected helped defray expenses of the voyage, and investors reaped dividends in proportion to its success. Therein lay the gamble. If a ship fared badly, investors lost in proportion.

One day during the busy epoch of American whaling there came to the port of Sag Harbor on Long Island an enterprising young man seeking his fortune. Whatever money he could afford to gamble in business he put into shares in an upcoming voyage of the whaler *Union*. He took lodging in town and settled down to a long, patient wait until the *Union* should return with a profit on his investment. 'Twas said that the young man whiled away the weeks and months by industriously writing stories in his quarters.

---

*It was when I joined John Steinbeck as a judge in the Sag Harbor Whaling Festival that I first met the great novelist. Through our common interests in the sea, sport fishing, and writing we became good friends. Together we enjoyed offshore bouts with tuna, sharks, and other finned game, and memorable are our long discussions of the ways of sea creatures. My warmest memory of John Steinbeck, though, is that of a wonderful companion. I'd like to dedicate "Whale Tales" to that memory.

The *Union* remained at sea for nearly three years, but the voyage was a financial fiasco. Our young hero literally lost his shorts.

It's to be wondered what would have happened if the *Union*'s voyage had been highly profitable. Local history has it that the industrious young man was James Fenimore Cooper.

A century later history repeated itself, with a couple of variations. There came to Sag Harbor another writer, also industrious but with a dramatic difference. This writer's name was known around the world—John Steinbeck. John and his wife Elaine established a spring-summer-autumn home in Sag Harbor, and there they lived until he died. With his love for the sea, it was natural that John should add Sag Harbor's whaling history to his interests (others of which were marine ecology and biology, and ichthyology). He was a founder and leading mentor of the village's Whaling Festival, an annual event commemorating Sag Harbor's romantic past.

Much Given, a Lot Taken

Submarines are potent war weapons, all the more so for being cloaked in shadows of the deep. But they also suffer a substantial mortality rate in combat, now increased by the alert "eye" of sonar.

During World War II some fifty-two U.S. subs were combat casualties. With them went 374 officers and 3,131 men.

Statistics on the exact maximum size of Japan's undersea boat fleet in that conflict are not readily available (sources in Tokyo do not answer inquiries). But its recorded losses were calculated at 128; and many of the fifty-eight Japanese subs left at the end of the war were so much junk, unable to operate.

In both World War I and World War II the Germans placed a heavy value on U-boats. During the latter Armageddon they produced them like batches of cookies. Their subs relentlessly prowled everywhere in search of prey, often in dreaded "wolf packs." An idea of how important the Nazis considered U-boat warfare is seen from the fact that in January of 1942 they had ninety-two operational, and although eighty-seven were lost that year, new construction had 212 in service by December.

From early on in World War II German submarines wreaked fearful destruction. Two ace skippers, Prien and Kretschmer, were credited with the stunning total of 200,000 tons of enemy shipping *apiece*. During 1942's early months the U-boats found so many targets off the U.S. East Coast, among convoys and in the Caribbean, that their commanders dubbed the period "Happy Times." Within six months twenty-one German subs sank 500 Allied ships.

At one stage it appeared that U-boats would win the war for the Third Reich simply by throttling the Allies' life lines at sea. And they might well have done it, too, if their avowed objective of 800,000 tons *per month* (realized at one point in World War I) had been attained. Closest they came was 650,000 tons in one month, not exactly something to sneer at.

But this wholesale extermination was slated for collapse, fortunately. The Allies knocked it into a rapid downhill course with growing numbers of aircraft and war vessels, improved weaponry, and development of a better escorted convoy system. Germany's destruction of enemy shipping began to shrink, and the Allies' elimination of U-boats increased. During one four-month period 109 Nazi subs were sent to the bottom.

It's hair-raising when you realize how close to U.S. shores German submarines ventured duringWorld War II. Many instances we'll never know, but I can cite a couple.

Best known was that occasion when one put a group of saboteurs ashore at Napeague on the southern coast of Long Island. Thanks to the alertness of a Coast Guard beach patrolman—among the great heroes of the war, in my opinion—they were caught before they could execute their plans.

On May 5, 1945 the German sub U-853 sank the collier *Black Point* at the entrance to Long Island Sound. That same day, U-853 was erased by the destroyer escort U.S.S. *Atherton* and U.S. Coast Guard frigate *Moberly*.

A friend of mine, former party boat skipper Russ Redfield of Freeport, New York, told about a frightening incident while serving aboard a Coast Guard patrol vessel. One moonless night, just off Long Island's South Shore, Russ said, they spotted a dark silhouette slip between them and the beach. With night glasses they made out the profile of a surfaced U-boat. Their patrol craft was not equipped to engage the sub, but the raider's presence was no longer a secret.

"Homing Instinct"

It was a fine day in 1849 when the sailing ship *Minerva* stood out from Ely's Harbor, Bermuda, and as a spanking breeze filled her canvas the captain set a course for distant ports in Africa and the Orient.

Minerva never reached port, any port. After months passed without sign or word of her, vessel and crew were written off as lost. Long voyages were calculated risks in those days, so no one was astonished when a ship turned up missing.

But one day in 1851 the citizens of Ely's Harbor *were* astonished. In port again after a long absence was a familiar vessel, the *Minerva*. Her sudden reappearance was strange in more ways than one. Grounded on the harbor's shoals, she was a far cry from the trim craft last seen two years earlier. She looked much the worse for wear, a forlorn derelict.

A boarding party found her to be completely devoid of human life, and there were neither corpses nor skeletons. The last entry in her log had been written a year before. There were no clues to the crew's fate . . . nor was it ever learned what happened to them. But by an eerie twist of destiny—or was it under the guidance of a phantom crew?—she found her way across thousands of miles of open sea from the Indian Ocean to her home port.

A Clouded Crystal Ball

During a blinding snowstorm on February 17, 1893, the northbound schooner *Elsie Fay* ran hard onto some rocks at Montauk Point, New York. As her crew of seven prepared to abandon the stricken 172-ton vessel, her mascot, a talkative parrot, reportedly screeched, "We'll all go to hell together, boys!"

Fortunately for the crew, but unhappily for the loquacious bird, that prediction was slightly in error. The parrot was the only casualty, found frozen stiff on the beach the next day.

At the time of her accident the *Elsie Fay* was bound for Boston from the West Indies with a large quantity of coconuts in her holds. Under battering by waves she broke up, liberating thousands of coconuts that washed up on the beach to be gratefully collected by local residents. People thereabouts ate coconut meat prepared in just about every conceivable way for months afterward. From that day to this, the section of beach where the schooner was wrecked has been called The Coconuts, and you'll see it labeled as such on surf fishermen's maps of Montauk Point.

Monarchs of the Seas

In their 1800s epoch passenger ships such as the *President* and *Great Eastern* were considered leviathans. How their beholders' eyes would have popped if they could have seen some of the monstrous vessels of the decades from World War II on! Consider ships such as the following.

The U.S. aircraft carrier *Enterprise*, the "Big E," is 1101 1/2 feet long and nuclear-powered. The carrier *Admiral Nimitz* is 65 ½ feet shorter than the Big E but the warship had the world's greatest full-load displacement, 95,100 tons. It's reported that the nuclear-powered *Admiral Nimitz* can cruise for thirteen years without refueling. Assuming twenty-five years as an average life for such a ship, she would have to refuel only once in her entire service career.

The super-supertankers of the Universe Class, built in Japan, with a dead-weight tonnage of 326,500, are 1,135 feet long with a 175-foot beam. Among the super-giants are the *Globtik London*, 1,242 feet 10 inches long, 483,939 dead-weight tons, and Shell tankers reported at 540,000 tons.

The *France* is the longest and costliest liner ever. She measures 1,035 feet long overall, and her price tag came to a whopping $81,250,000. (For contrast, constructing and launching the 693-foot *Great Eastern*, largest ship afloat in the mid-1800s and for nearly half a century, cost between $5,000,000 and $6,000,000.) The *France* barely squeaked by the old *Queen Elizabeth* in length, topping the magnificent British liner by only 4.2 feet; but the *QE I* outweighed the French Line's monarch, 83,673 gross tons to 66,348. Cunard's old *Queen Mary* was no slouch in the size department either. Built in the 1930s, her specifications included a length of 1,019.5 feet and a gross tonnage of 81,237.

Here are the largest battleships ever to threaten enemies. Distinctions in this department are claimed by Japan and are of World War II vintage. They were the titanic "twins," *Yamato* and *Musashi*, each 863 feet long, 127 feet wide, with a full-load displacement of 72,809 tons. They also claimed the largest guns ever mounted on warships—18.1-inch caliber, 75 feet long, hurling a 3,200-pound projectile. Despite their size and weaponry, however, these "battle wagons" proved to be as mortal as any other warships. Bombs and torpedoes sent the *Musashi* to the bottom of the Philippine Sea in a battle on October 24, 1944. U.S. planes similarly dispatched the *Yamato* in Bungo Strait on April 7, 1945.

Poetic Justice, Maybe

Many a vessel has been sent to the bottom by natural hazards and war, but rarely can a fish claim such a victory. One obscure historical record is the case of the schooner *Red Hot*, permanently chilled by a large game fish in the 1890s. According to a fragmentary account, an enraged swordfish charged the schooner, stove in her planks, and sent her to the bottom off Fire Island, New York.

There was no mention of what the *Red Hot* did to vex the big billfish. On the basis of intimate knowledge of swordfish and firsthand angling encounters with them, my guess would be that the schooner had wounded the beast with a harpoon. I don't believe I've ever heard of an *unprovoked* attack by a swordfish.

But I don't doubt for a moment that a full-grown swordfish, with a weight potential up to 500 pounds and more (even heavier in the Pacific Ocean), could sink a wooden vessel, even a small schooner. I personally know of an instance in which a white marlin, much smaller than a swordfish, drove its bill or spear completely through the side of a wooden sport fishing cruiser. Years ago, noted American naturalist Dr. William Hornaday told of a swordfish that slammed its weapon through a ship's plank that was of seasoned oak, three inches thick and copper-sheathed. Not only was the sword driven through to the hilt, the force was so great it appeared that it had been broken off. It didn't sink the vessel. The imbedded sword was discovered later when the vessel was out of water for repairs.

Guts

In our age of shooting men into outer space we easily forget the enormous courage demanded of ancient mariners who undertook transoceanic voyages in sail-powered—even oar-powered—vessels that were mere cockleshells, some not much larger than cruisers used in deep-sea sport fishing today. Like outer-space explorers, they faced many unknown quantities. There were unpredictable weather and water conditions. They also faced fancied terrors, such as hideous sea monsters, then believed to be lying in wait way out yonder. Like astronauts, they never could be certain that they'd get back safely. What's more, they did it all without any means of communication.

Such an epic in courage was the voyage of Leif Ericson, son of Eric the Red, and his men. According to Icelandic sagas, those Nordic adventurers sailed westward across the Atlantic Ocean circa 1000 A.D. and discovered a land Ericson called Vinland because of grapevines he found there. Historians have variously identified his Vinland as Labrador, Newfoundland, and New England.

Nearly 950 years later, a bold little band composed of a Captain Folgero and three companions sailed a facsimile of Leif Ericson's Viking vessel from Korgen, Norway, to Boston, Massachusetts. The "retracing" consumed three months and twenty-four days.

Hold On!

There's a limit to which vessels can roll, or lean, and still recover a normal upright position. Part of ships' instrumentation is a device called an inclinometer, whose function is to indicate the degree of roll.

Destroyers, affectionately nicknamed "tin cans" by navy men, have an unenviable reputation for rolling like mad in heavy seas. They can be among the worst ships in this respect. But even they outdid themselves during an incredibly fierce typhoon that clobbered the U.S. Navy's Third Fleet in the western Pacific on December 17 and 18, 1944. This typhoon was notable on several counts in addition to its ferocity. The barometer plummeted seven points in a single hour, reaching an unheard-of low of 27.30. Winds attained velocities so high they could no longer be measured. They were estimated to be in excess of 110 knots, better than 125 miles an hour. And towering waves were like toppling mountains.

At the peak of this fury the destroyer *Dewey*, DD 349, set some sort of record for rolling. At 1006 hours (10:06 A.M.) the inclinometer on her bridge showed rolls through forty to fifty degrees. At 1130 hours it indicated a roll of seventy-three degrees to starboard (her engine-room inclinometer showed seventy-five) where she paused for a few seconds as though debating whether or not to go all the way. She couldn't have come much closer to flipping over. At ninety degrees her stacks would have been parallel to the sea's surface. The *Dewey* survived, but she was almost mortally crippled.

The toll of that typhoon among the Third Fleet was as terrible as its fury. These were just some of the items on the bill: 790 men dead or missing; upwards of eighty injured, many seriously; loss of the destroyers *Hull*, *Spence*, and *Monaghan*; 146 aircraft blown overboard or damaged beyond repair; major damage to at least eighteen ships, including cruisers, light carriers, destroyers and destroyer escorts; and varying degrees of repairs needed by battleships and heavy carriers.

Actually there's a greater roll than the *Dewey*'s on record, but it has to be qualified. Amid rough seas off the Oregon coast in November 1971, the U.S. Coast Guard motor lifeboat *Intrepid* turned turtle completely, through 360 degrees, and righted herself. This "record" has to be qualified because Coast Guard motor lifeboats are designed to be able to recover when they flip over, thanks to an extremely heavy keel and a hull that doesn't ship a lot of water.

For a Clean Sweep

Perhaps you've wondered why naval vessels display on a mast an object so incongruous as a broom after victory in battle or other notable accomplishment.

Legend has it that the custom was born in the seventeenth century. It started when Dutch admiral Marten Tromp tied a whip to his mast to support his claim that he could beat the British. When English admiral Robert Blake defeated Tromp's fleet in their next encounter, he tied a broom to his flagship's mast to indicate that he had swept the Dutch from the seas.

Since then, a broom so carried aloft symbolizes a "clean sweep," or a job well done.

A "Candle" for the Nation's 200th Birthday

Over a span of nearly three centuries millions of fishermen and inbound and outbound ocean travelers have seen Montauk Lighthouse, a tall sentinel on a rise the Indians called Turtle Hill at the easternmost tip of Long Island. Generations of mariners have been guided by its Cyclops eye of light in the gloom of night and heard its basso profundo horn in fog. But relatively few viewers are aware of this structure's venerable significance.

President George Washington himself authorized construction of Montauk Lighthouse on August 18, 1795. Costing approximately $22,000 to build, it still stands atop Turtle Hill. Beneath its bright eye have passed just about every conceivable kind of vessel known to man from the American Revolution era onward: little fishing smacks to squareriggers; U.S. warships, early-vintage types under canvas to nuclear-powered submarines; passenger ships, schooners, and clippers to the largest luxury liners; and sport- and commercial-fishing vessels, humble outboard boats and expensive cruisers to big trawlers.

For uncountable thousands of foreigners visiting the United States by ships (by some airliners too) Montauk Lighthouse has been their first glimpse of this country. For numerous Americans returning from abroad its message has been "Welcome home!"

I can't think of a better candle for the nation's 200th birthday cake.

23

Death from Outer Space?

The year was 1948. The month was February, summer in the Southern Hemisphere. Clear, warm weather smiled on the Dutch freighter *Ourang Medan* as she plodded through a calm sea in the Straits of Malacca between Malaya and Sumatra en route to her next port of call, Jakarta in Indonesia.

Abruptly other ships and shoreside listening posts became aware of the *Ourang Medan*'s presence by repeated distress calls accompanied by mention of her position. Ships in her general area were alerted and rescue vessels put out from coastal locations. As abruptly as they began, her distress calls ceased with a jolting final message. The captain and all the *Ourang Medan*'s officers were dead, it reported, adding that probably the entire crew was dead. Then followed an undecipherable portion. The last words from the freighter's radio operator were distinct: "I die."

Aided by radio direction-finding equipment, vessels bent on rescue pinpointed the *Ourang Medan*'s position and found her within a matter of hours. From a distance there didn't appear to be anything strange about the freighter. Smoke trailed from her funnel, although she seemed more to be drifting than under power. Her radio was silent. No one could be seen moving about her decks, as should have been the case if she were in sufficient trouble to necessitate an SOS. Nor was there any response to hails from the approaching vessels.

A Stunner

Investigating parties boarded the *Ourang Medan* and confirmed her radio operator's startling announcement. All hands, captain to lowest-ranking seaman, were indeed dead, their corpses sprawled on deck and in various other locations, including the wheelhouse and chart room. The skipper's body measured its length on the bridge. Her radioman, true to his final words, "I die," was slumped lifeless at his post, a hand still on his instrument's transmitter key. Whatever calamity struck the *Ourang Medan*, it was thorough. Even her canine mascot was dead.

Such complete mass destruction in itself is extremely peculiar,

176

but what makes it all the more unfathomable is the reported description of the corpses' attitudes when discovered. According to the boarding parties' report, all their faces were turned toward the sun. To the man, their eyes were staring and their mouths gaped widely. The ship's dog had died with his teeth bared in a snarl. From these details it could be surmised that all the deceased, including the dog, saw what was about to destroy them and died in terror.

The *Ourang Medan* too was about to die, and it was almost as though she were conspiring to keep the secret to her crew's fate. Just as the boarding parties had decided to tow the death ship to port for thorough investigation, fire suddenly erupted in one of her holds and threatened to envelop the entire freighter. Flames rapidly spread beyond control, and the boarders had to flee for their lives. It was good they did. Minutes later the doomed ship was torn by an explosion. In sheets of flame and clouds of smoke and steam she rolled over and quickly sank into the depths of the Straits of Malacca.

When the whirlpools of her death throes closed over the *Ourang Medan* they locked the true story of her crew in the files of the sea's unsolved puzzles.

Again, a Guessing Game

There are theories about the cause (or causes) of the total demise of the *Ourang Medan*'s crew. In unsolved mysteries there are always theories. The trouble is, any which seem to have some basis in logic do not fit into the picture satisfactorily.

Epidemic disease, for example. This could have been verified or rejected positively only by autopsies. There were no bodies to autopsy. However, logic weakens the disease theory. It's possible that there exists an infectious, extremely virulent disease, perhaps very rare and as yet unrecognized by medical science, that can *mass-kill quickly*. But we have to base our logic on present knowledge, which does not include a disease with such a pattern. Besides, the ship's dog was dead too, ostensibly from the same cause. So far as I know, diseases affecting humans generally do not bother animals—and vice versa.

Poison or contaminated food are possibilities, albeit unlikely ones. Who could or would administer poison to an entire crew—and a dog—simultaneously? And even if an assassin could, who was its administrator? Botulism E, the most deadly food poisoning known to medical science, is almost invariably fatal without emergency treatment and kills within hours. But again it's highly unlikely that so many victims could have been affected lethally at the same time. Further, there apparently was no mention of illness by the radio operator in his distress calls. Again, we'll never know, lacking post mortems.

Asphyxiation has been considered. It can kill *en masse* and rather quickly. Witness the wiping-out of families by carbon monoxide seeping from a faulty heating system. Perhaps there were lethal gases, such as CO, aboard the *Ourang Medan*, or fumes from a fire already smoldering in a hold. The big flaw in an asphyxiation theory is that gases or fumes probably would not have killed people out in the open air, and certainly not all of them. Too, it has been pointed out that humans and animals killed by carbon monoxide or other lethal fumes usually die with their eyes closed, as though peacefully asleep. The men and dog on the freighter were found with their eyes and mouths open in attitudes that looked anything but peaceful.

Some unofficial analysts of the *Ourang Medan* case have added another note of mystery by viewing the "sudden" fire with suspicion. No mention of a fire was made in the radio operator's last message, so we can only suppose that it started—or was discovered—sometime during intervening hours while rescue vessels were on their way. What seems very odd is that fire should break out suddenly so soon after boarding parties arrived on deck. An explosion soon after the boarding parties abandoned her also seems strange. The timing of these two developments must be viewed with suspicion. One is tempted to wonder if the fire, by whatever or whomever set, was calculated to discourage the would-be rescuers, and if the explosion was calculated to effectively remove all evidence. (If so, it must have removed any human conspirators too.)

While we're in a conjecturing mood we might as well go the full route and ask: Were the men aboard the *Ourang Medan* struck down by some unknown force from the sky—a UFO, perhaps? That is one chilling thought.

Like the cause (or causes) of the victims' death, the Dutch freighter's exact toll will never be known. Evidently there was no shoreside record of the total number of persons aboard, which conceivably could have included some passengers. And the boarding parties didn't make a full count of the corpses, which is understandable, considering plans to tow the ship to port, and the outbreak of fire. For the same reasons, presumably, no count was made of any dead below decks, and it was not determined if there was anyone below still alive. The *Ourang Medan* kept those details secret too.

24

Dives to Oblivion

French novelist Jules Verne lived during the Victorian decades, an epoch noted for its straight-laced policies in practically everything from conversation to sex. Perhaps that was one reason his books were enormously popular in Europe and America. They offered an escape.

Monsieur Verne was far, far ahead of his time. In fact, he probably was the first author to achieve international fame as a science fiction writer. His vivid, fertile imagination created all kinds of wonders for the future; and a number of his ideas amazingly became reality. For instance, he foresaw rocket travel to outer space in his book *From the Earth to the Moon*. And although the vehicle was a balloon, he foresaw long-range air travel in *The Tour of the World in Eighty Days*.

Jules Verne's keen imagination also sired a highly advanced version of the submarine. He enthralled readers with his *Twenty Thousand Leagues Under the Sea*, in which an idealistic but slightly mad Captain Nemo endeavors to curb conflicts between nations by indiscriminately sinking every war vessel he encounters with his fabulous underwater ship, the *Nautilus*. If you read that book or saw its marvelous interpretation as a motion picture by Walt Disney, and looked between the lines, you got an impression that maybe Jules Verne foresaw nuclear power too.

Nearly a century later, Jules Verne's *Nautilus* had real-life counterparts in the U.S. Navy's nuclear-powered submarines (including one with the same name). There are marked parallels between the fictional and actual undersea vessels: Size, extra long-range cruising capability without refueling, and impressive destructive power, all intended as a deterrent in case certain nations get overambitious.

There's a tragic parallel too. Like Verne's *Nautilus*, a couple of the modern submarines died prematurely at sea, taking their crews with them.

U.S.S. "Thresher"

An awesome war machine, the *Thresher* was rated at 4,300 tons (when submerged) and reportedly had speeds of twenty knots on the surface and twenty-five to thirty knots below it, and could travel up to 60,000 miles without refueling. After some adjustments in a Navy yard she put to sea in April of 1963 to occupy herself with deep-diving trials in the Atlantic off New England. Aboard were 129 men, including several civilian technicians. Accompanying her was a submarine rescue ship, and the two vessels maintained radio contact.

On the morning of April 10th, when about 220 miles east of Boston, *Thresher* announced that she was nearing her "deep-dive test depth," interpreted as a submersible's maximum depth limit. Nothing in this routine communiqué indicated anything wrong. But it was followed in rapid order by two messages. The first mentioned an "up angle" (presumably referring to the sub's diving planes) and attempts to blow the ballast tanks. It could be gathered from those details that she was trying to come up, or perhaps check further descent. Then came a sound like that of high-pressure air emptying the ballast tanks. The second message was too garbled to be understood and was drowned out by noises described as sounding like breakage. After that, ominous silence. The *Thresher* was quickly reported overdue and presumed missing.

The Navy concentrated all kinds of sophisticated looking and listening devices on an intensive sea-air search for the multimillion-dollar sub and her priceless human cargo. Surface vessels scoured square miles of Atlantic Ocean, seeking telltale oil slicks and flotsam such as life jackets and other buoyant articles that may have popped to the surface, as well as any emergency marker buoys the *Thresher* might have released. Aircraft fanned out to give their double advantage of height and speed. Sonar scanned the sea floor to locate structures, like a submarine, that normally wouldn't be there. The latest underwater listening gear provided sensitive ears to detect any unnatural sounds in the depths.

Nothing.

The 300-foot *Thresher* seemed to have taken a place on the roster of vessels whose real fate would never be known.

Few submersibles are capable of entering the sea's greater deeps, where water pressures increase to 2,000 or 3,000 or more pounds per square inch and could crush an ordinary submarine like an egg shell. But the U.S. Navy had a special submersible to send in search of the *Thresher*. It was the bathyscaphe *Trieste*, proven in dives to a fantastic 35,802 feet in the Pacific Ocean's Marianas Trench

180

(January 23, 1960, with Dr. J. Piccard and the Navy's Lt. D. Walsh aboard). After several dives the *Trieste* located the *Thresher*'s remains, dead at a depth of 8,400 feet.

At that depth the pressure is nearly 4,000 pounds per square inch.* The *Trieste*'s findings indicated that the ill-starred sub had imploded—that is, been crushed inward—under enormous pressure and was broken into pieces. Bathyscaphe observers said the scene on the ocean floor looked like a junkyard. It's better if we do not try to imagine what happened to her men.

At least two parts of the *Thresher* mystery were solved, the "what" and "where." Obviously she submerged beyond control. But an equally important question—"why"—had to remain unanswered. There is just no way at present in which a post mortem can be conducted at such depths; and salvage, if possible, would be very costly. Naval experts theorized that the failure or collapse of some vital fitting, or fittings, admitted sea to her hull, filling it. Unable to rise, the sub then would have dropped beyond her safe maximum depth into pressures causing implosion. Another theory offered was that something went wrong with the electric circuits of her controls, preventing her rising.

Thus did the *Thresher* join four other U.S. submarines—O-9, S-51, S-4, and *Squalus*** in death off New England.

Occasionally some good can come even from tragedy. Loss of the *Thresher* brought about revision of deep-diving procedures for nuclear submarines to increase safety.

The "Scorpion"

Lessons are learned, often the hard way, which cause equipment to be improved and procedures to be revised for safety in man's unceasing efforts to dominate the sea. But King Neptune's empire is still a largely unknown frontier, with unpredictable combinations of its own and man-made circumstances. He is just as determined as

---

*The weight of the atmosphere we carry around on our shoulders is about 14.9 pounds per square inch.

**The *Squalus* was only about two weeks old, but already had completed several dives without incident, when she made a routine submersion run on the morning of May 23, 1939. Suddenly water began pouring in through the main engine induction valve, flooding her after compartments. The *Squalus* settled to the bottom in forty fathoms of sea. Fortunately, at that depth it was possible to use a special rescue diving bell, and thirty-three of her crewmen were saved. In time the *Squalus* was raised, repaired and reconditioned, after which she re-entered U.S. Navy service as the *Sailfish*.

we are—maybe more so—to show who is boss. More frequently than not, that point is proved, despite even more sophisticated equipment and refined navigation procedures. Witness the loss of the U.S. submarine *Scorpion*, five years after the *Thresher*.

Like her tragic predecessor, the boat* *Scorpion* was nuclear-powered, but at 256 feet and 3,075 tons, she was somewhat smaller. *Scorpion* is an historic name in the U.S. Navy. The submarine became the fifth vessel to be so christened. Two of her ancestors were sailing ships, both combat casualties in the War of 1812. The other two were gunboats of later 1800s vintage.

Scorpion took to the water in a 1959 launching. There was nothing untoward in her performance, with the possible exception of some relatively minor "bugs" that show themselves in any new piece of machinery. During the summer of 1967 she was drydocked to square away a few problems—nothing especially serious, apparently—and that October she underwent new sea trials with ninety-nine officers and crewmen aboard.

Scorpion was due at the U.S. Navy's complex in Norfolk, Virginia, from a sail in the Mediterranean in the afternoon of May 27, 1968. Her radio had last been heard on the 21st while she cruised homeward off the Azores. At that point her position was approximately 2,300 miles from Norfolk, and seemingly all was well. There were no further messages from her, but neither was there cause for concern—yet.

The undersea boat didn't reach Norfolk at her ETA. Now the lack of further communiqués from her took on significance, and she was declared overdue as of 1300 (1:00 P.M.), May 27th. A large-scale search was instituted without delay, in the air, on the surface and below the waves. Planes, destroyers and submarines systematically scouted the Atlantic off Norfolk and to the east in a pattern blanketing more than 100,000 square miles. They found nothing.

During the probing it was reported that a vessel, presumably a freighter, had spotted an oil slick nearly 600 miles east of Norfolk along about the 23rd of May, four days prior to the *Scorpion*'s anticipated arrival in port. Aircraft and ships were dispatched to that area. One of the search vessels, U.S.S. *Hyades*, sighted an orange buoylike object bobbing and drifting free on the surface. It was an item that could be identified as part of a submarine's gear, but it turned out to be only a loose article. Air and surface coverage of the region failed to sight an oil slick such as had been reported. Then

*If there's anything that will make a seafarer wince, it's to hear a ship called a "boat." Yet in naval parlance it's quite proper to refer to a submarine as such. Here "boat" may be a holdover from "undersea boat," an early term.

weather and sea moods turned ugly, hampering further search. Nothing definitely identifiable as a clue had been found.

Days passed. For a brief interval searchers thought they might have located the *Scorpion* when the wreck of a submarine was discovered on the bottom off Virginia. This corpse subsequently turned out to be of World War II vintage, nationality not mentioned. Now the search shifted to the Azores and the region from which *Scorpion* had last reported. Here the results were equally disappointing.

Finally ships were withdrawn from the search, but two U.S. oceanographic vessels, *Bowditch* and *Mizar*, remained in the area. The *Mizar* probed the ocean floor with a special sled outfitted with an undersea camera and powerful lights. This highly specialized device was towed at a very slow rate, about one knot (roughly 1 1/8 land miles per hour), just over the sea bottom. On October 29th its camera lens found the *Scorpion*.

Photographs showed the dead sub reposing on the Atlantic's floor well off the Azores and nearly two miles down. There was no doubt. Identifiable details could be made out. Study of one picture revealed a length of line dangling from an emergency buoy compartment. Ostensibly it had been attached to that orange buoylike object sighted by the *Hyades*. The camera's eye also captured views of parts and sections of the boat lying some distance away from her hull. Recalling the "junkyard" appearance of the *Thresher* scene, and adding the fact that *Scorpion* lay in even deeper water where pressures were greater, a logical conclusion was that implosion was the demolishing agent here too.

Again we have the "what" and "where" answered, but are without a reply to the "why" or "how."

An earlier letter home from one of her crewmen mentioned the *Scorpion* being in some sort of chance encounter with a Russian destroyer on the Mediterranean about eleven days prior to her last communiqué when off the Azores. Although the "tête-à-tête" seems to have been anything but amiable (the destroyer reportedly had her guns trained on the sub) there was nothing to link the Soviet warship with loss of the *Scorpion*. The possibility of a connection did not escape consideration, however.

Also part of the story is an eerie radio message intercepted by several vessels two days *after* the missing submarine was declared overdue in Norfolk. It read: ANY STATION THIS NETWORK, THIS IS . . . [the Scorpion's code name was given]. The message was branded a cruel hoax, but I've since pondered a question: How did the canard's perpetrator know the sub's code name?

In light of the *Thresher* disaster and the similar implosive appear-

183

ance of the two subs' remains, a logical theory was that *Scorpion* met a like demise: That is, some sort of malfunction—mechanical, electrical or perhaps even human, or a combination of same—resulted in uncontrollable flooding of the hull, sinking the boat far beyond her safe maximum depth, as in the *Thresher*'s case. Also conjectured was the possibility of an accident involving a torpedo. One of those lethal projectiles certainly could kill a submarine and scatter some wreckage, but the implosion appearance of the *Scorpion*'s remains would seem to rule out the torpedo theory.

In November 1968, the public learned that a sea-air search involving upwards of 5,000 men and some 400 vessels and aircraft had culminated with the discovery of what was left of the *Scorpion*. What couldn't be announced with any certainty was the cause of a disaster that carried ninety-nine members of Uncle Sam's navy and a costly fighting machine to their doom 10,000 feet down.

The Guiness Book of World Records lists the *Thresher* as the worst submarine disaster in history.

25

Anyone Hear from Josh Slocum?

"It's just to save buying a winter overcoat," chuckled Captain Joshua Slocum when friends asked him why he sailed alone to the West Indies every autumn. The world-famous skipper was being facetious, of course. Actually, in his sunset years he wearied of New England winters: and, if the truth were known, he probably got restless for the open sea along about that time too. Whatever the motivation, it became Captain Slocum's custom to set out on a solo voyage in his beloved *Spray* to the Caribbean each fall. For skipper and boat—which, in different phases of her long career, was rigged first as a sloop, then as a yawl—it was a routine run. Captain Joshua Slocum was a man of enormous seafaring experience. Breezing down to the Caribbean alone in a thirty-six-foot sailboat was about as difficult for him as paddling a canoe across a small lake is for a summer vacationer.

For this particular trip in the autumn of 1909 the *Spray* had been refitted and all spruced up at a yard in Bristol, Rhode Island. Captain Slocum, pushing age sixty-six, was in fine fettle and raring to go. The venerable, inseparable pair was off to Grand Cayman Island in the Caribbean Sea.

Somewhere between Bristol and Grand Cayman these inseparable companions, veterans of amazing adventures all over the world, vanished forever, becoming still another secret in Davy Jones' Locker.

The disappearance of Captain Joshua Slocum and his *Spray* is not a mystery *per se*. Obviously, something fatal caught up with them. It's the nature of what overtook them that remains an enigma. And to understand why, you must know details of the fabulous mariner's background. His story is in classic seafarers' tradition. Robert Louis Stevenson would have written a book about him.

Sea Water, Not Blood, Ran in His Veins

Joshua Slocum, master mariner, began his life in Wilmot, Nova Scotia, in 1844, son of a farmer. As early as eight he was working on the family farm between bouts with school books. And when his father exchanged the drudgery of agriculture for a bootshop in the bayside hamlet of Westport two years later, the boy became an apprentice bootmaker. There are reasons to believe that this occupation enchanted him even less than piloting a horse-drawn harrow.

His father being very thrifty, Josh's schooling was terminated at age ten so he could put in full time in the bootshop. He hated it cordially, and found escape in gazing out at Westport harbor and yearning to be aboard tall ships that came and went. For two years he endured the detested work, bolstering his tattered morale by daydreaming about when he could go to sea. Finally relations between him and his strict father deteriorated to the point where they became intolerable. Josh translated his daydreams into reality and ran away to sea at age twelve. Unhappily for the ship-struck lad, the adventure was aborted, and Josh received a sound thrashing.

A later attempt at escape was successful. Josh got himself a berth on a St. Mary's Bay fishing smack as—of all things for one so young and inexperienced—a cook. Thus did Joshua Slocum and the sea become wedded forever.

Josh was a dreamer, yet practical. Working aboard sailing vessels in Nova Scotian waters got his feet on a deck and away from hated shoreside jobs, but he ached to explore far horizons. It mattered little to him that his first long stride toward those horizons was aboard a ''deal droger'' headed for Ireland. The destination was fine; the vehicle was not. In regional mariners' language, ''deal'' signified a cargo of lumber and ''droger'' was a contemptuous term for a leaky, ill-found vessel, usually with poor chow for her crew. Josh signed aboard a Dublin-bound lumber carrier, a slow, awkward ark of a ship whose mate periodically had to bawl ''Man the pumps!'' to keep her afloat.

Josh left the droger in Dublin and took a packet for Liverpool. That old English port turned out to be his threshold on the world. From then on, voyages aboard tall ships carried him to Africa, China, the East Indies, Australia and exotic island ports in remote reaches of the globe. Along the way he taught himself navigation, an education that was to stand him in good stead the rest of his life.

As years passed he worked his way diligently up through the ranks, and then one day he earned his master's papers. As captain he commanded several vessels, among which were the trader *Amethyst*, the bark *Washington*, the passenger packet *B. Aymar*, the barkentine *Constitution* (aboard which his son Victor was born), and another trading ship, the *Pato*. He logged tens of thousands of miles in transpacific service between U.S. West Coast ports and the Far East, Philippines, Australia and the South Seas. And those were not all of his horizons. With his command of the 233-foot, 2,000-ton clipper *Northern Light* and the 140-foot, 365-ton bark *Aquid-*

neck he expanded his travels even further by voyages to Europe and South America. In some of these ships his interest was not only their captaincy, but also part or whole ownership.

All these voyages turned out to be groundwork for his supreme adventure, sailing alone around the world in a 36-foot craft, a voyage that was to perpetuate his name in U.S. maritime history.

Preparations

Preparation began with reconstruction of an elderly sloop named *Spray*. She belonged to one Eben Pierce, a New Bedford whaling captain, and lay forlorn and neglected in a pasture outside Oxford Village, Massachusetts, near Fairhaven, where Captain Slocum had taken up residence. Although the aged sloop was of dubious ancestry, and even more doubtful health, she appealed strongly to Josh. In her he saw a vessel such as a man might sail around the world. He acquired her.

Mainly he was interested in her lines; and with the boat serving as her own template he set about extensive rehabilitation. Actually this was near-complete rebuilding, but with preservation of her original lines. One alteration he did effect was elimination of her centerboard, whose well Captain Slocum believed weakened a hull. Part of *Spray*'s new body was a keel hand-hewn with a broadax from a single piece of pasture oak. Pasture oak was white oak, the nickname "pasture" stemming from the fact that it was cut from a tree that had grown up all by itself out in a field. New England small-craft builders favored pasture oak over forest oak because it was tougher, having had to withstand the elements and four winds to survive. Pasture oak made an exceptionally strong, durable keel, frames, stem and sternpost. Steam-bent, good and hot, it could be fashioned into any desirable curve or angle. When cold it got hard as flint. Captain Slocum fashioned *Spray*'s new planks from 1 1/2-inch Georgia pine.

Rebuilding completed, the *Spray* was thirty-six feet, nine inches long overall, with a fourteen-foot two-inch beam and an inside depth of four feet two inches. Her layout included an anchor locker and stowage forward, a tiny galley just abaft her mast, then a roomier cabin, and finally a small cockpit with wheel steering. There was no engine. Her gross tonnage was 12.71. *Spray* began her partnership with Captain Slocum as a gaff-rigged sloop, with a long bowsprit and a mast of live New England spruce. (Later he changed her rigging to that of a yawl. Four sails gave a total of 1,161 square feet.) And she emerged from his loving rehabilitation as a stable, responsive, very sturdy, extremely seaworthy boat. She served him well until death parted them.

187

The *Spray* was ready for her greatest adventure, a venture about which Josh Slocum had dreamed for those many years. He would sail single-handedly around the world.

The monumental voyage came in 1895 and began with a shakedown segment from Boston to Yarmouth, Nova Scotia, selected as her port of departure. Everything was in readiness on July 1st. Well provisioned and with her skipper in high spirits, the *Spray* stood out to sea from Yarmouth in a spanking nor'west breeze. Nine days later she logged 150 miles in twenty-four hours and was knifing water 1,500 miles east of Canada's Cape Sable.

What followed was an epic voyage in *any* vessel. In a 36-foot sailboat, solo, it was incredible.

After traversing the Atlantic, *Spray* departed Gibraltar for the Canary Islands. Originally Captain Slocum had intended to negotiate the Mediterranean to reach the Indian Ocean via the Suez Canal. He was discouraged from this by a warning that if he sailed along Africa's northern coast he might be captured by Berber pirates, then very active—and very ferocious. Josh altered his itinerary, but very nearly fell into pirates' hands anyway. Not far off the coast of Morocco a felucca set out in pursuit of the *Spray* and at one point drew so close that Captain Slocum could make out the faces of the craft's cutthroats. In their outfits they may have looked like bundles of old clothes on their way to a beggar's bazaar, but Josh knew they were vicious, murderous bandits. For once a gale was welcome. It saved him from the sea robbers.

Alternately battling squalls and creeping through doldrums after a landfall in the Capt Verde Islands, the boat picked up the sou'east trade winds. In September she crossed the Equator at 29.3° W. longitude. Propelled briskly by the trades, she fetched Pernambuco on the 5th of October, forty days out of Gibraltar. On she sailed: Rio de Janeiro on November 28th, then Montevideo, Uruguay. There she underwent repairs of some storm damage and on Christmas Day, 1895, she was ready to take on the perils of Tierra del Fuego and Cape Horn.

Spray made it around the Horn, but only by a whisker. A miserable region noted for its cold, wet weather, fog, fierce winds and treacherous currents, Cape Horn was the assassinator of many a sailing vessel attempting to circumnavigate it. For anxious hours at a time it was a tossup as to whether or not the *Spray* would become another victim. She sailed in and out of heavy weather, was pursued by inhospitable-looking natives, and was battered by williwaws, extremely vicious squalls known to roll full-size sailing ships on their beam ends. It's a gigantic tribute to Joshua Slocum's courage and seamanship and *Spray*'s performance that they survived some of the worst conditions the Cape Horn region could throw at them.

Violent torment didn't end with Cape Horn. Ahead lay a nightmarish place with a poetic, apt but disarming name, the Milky Way. Here, northwest of the Horn, the ocean turns milky white from the foam of huge seas churning and breaking among submerged rocks. The Milky Way was considered one of the worst death traps for ships on the face of the globe, and the *Spray* made it by night!

At the time, the only other vessel known to have survived the Milky Way was H.M.S. *Beagle*. Sixty years earlier, famed naturalist Charles Darwin viewed the hellish scene from the *Beagle*'s deck and wrote in his journal, "Any landsman seeing the Milky Way would dream of shipwreck and disaster for a week." Later Captain Slocum was to state that navigating the Milky Way became the greatest sea adventure of his life, adding that only The Almighty knew how he survived it.

By the time *Spray* cleared Cape Pillar, grim western sentinel of the Horn, she had logged 9,600 miles.

Out Across the Pacific

Now the course lay north and westward off the seaboard of South America. Word of the voyage already was preceding them and spreading Captain Slocum's fame. This he was to learn would happen with increasing frequency, port after port.

From Cape Pillar he steered for the Juan Fernandez Islands, some 340 miles off the coast of Chile. There Josh paid his respects to Mas-a-Tierra, the same little island on which Alexander Selkirk, real-life inspiration for Daniel Defoe's fictitious Robinson Crusoe, lived by his wits in solitary exile nearly two centuries earlier.

For the next forty-three days the intrepid pair alternately enjoyed fair weather and ran with a bone in the 36-footer's teeth through gales before fetching the Marquesas, nearly 4,000 miles from Mas-a-Tierra. Periodically sharks circled the *Spray*, as though they knew something Josh didn't; and once she came perilously close to collisions with surfacing whales as she glided into a gam of them in the darkness.

On July 16, 1896, a little over a year after leaving Yarmouth, the *Spray* dropped anchor at Apia in Western Samoa.

From Samoa the itinerary traced a course north of the Fiji Islands and south of New Caledonia, a place that would figure in world war news nearly half a century later. The *Spray*'s arrival in Newcastle, Australia, was theatrical—in a severe gale. Behind her lay a 42-day passage from Apia. On October 10, 1896, she rode the approach of summer Down Under into Sydney harbor. Sydney was a port with many memories for Captain Slocum, from his years as master of tall ships. Word of the voyage had long since reached the big Australian port, and he was accorded a hearty welcome that included a visit from officers of the British warship H.M.S. *Orlando*. The *Spray*'s recess in Sydney was so pleasant that she didn't leave until December 6th.

Onward, Ever Onward

Consulting his charts, Captain Slocum planned to sail south, then west, across the vast South Australian Basin, past stormy Cape Leeuwin projecting from the island-continent's southwestern corner, then on to the island of Mauritius in the Seychelles group, a lonely archipelago in the middle of the Indian Ocean. Via this route in the lower latitudes, he calculated, "westing" would be appreciably shorter.

Again, though, there had to be a radical change of plans. While dawdling pleasantly in Melbourne he learned that masses of ice were drifting northward from Antarctica into his planned route. This potential hazard, coupled with Cape Leeuwin's reputation for bad storms, turned him to other charts. He made the relatively short haul from Melbourne to Tasmania, and traveled about that island as he awaited favorable winds to propel him northward along Australia's eastern coast. The new route lay through the Coral Sea, along the 1,250-mile length of the Great Barrier Reef. Then he would steer west'ard to Thursday Island in Torres Strait, a turbulent flow between Australia and New Guinea. Thursday Island was 2,300 miles away.

En route northward he again stopped in Sydney, April 22, 1897, only for a brief stay this time. By now Captain Slocum was as used to storms as he ever would be, but one at Port Macquarie left an especially sour memory. In the blow he encountered a cutter yacht named *Guinevere* in difficulty, with three men—woefully incompetent, it turned out—on board. With problems of his own, Captain Slocum risked his neck to bring the *Spray* close to the distressed cutter, only to have his offer of a tow into port indignantly refused. It must have given him pause for thought about helping one's fellows, but he derived no satisfaction from hearing later that the *Guinevere* was lost—her crew saved, fortunately.

Any monotony of the northward segment along the Great Barrier Reef was broken by a few pauses, which included a stop in Cookstown, named for the great British navigator-explorer Captain James Cook. Off Cape Clarement came a change of pace that skipper and boat could have done without. The *Spray* came out second best in a brush with a coral head. Luckily, her stem wasn't damaged severely.

Reconditioned and reprovisioned, *Spray* left Thursday Island in Torres Strait on June 24th, and in a fresh trade wind scooted across the Arafura Sea for the Indian Ocean and her distant target beyond, Madagascar on the African coast. Now her speed was increased by the addition of a spinnaker, made for the skipper by a young lady back in the Juan Fernandez Islands, and set on bamboo given to him by Robert Louis Stevenson's widow in Samoa.

The *Spray* had been away two years and ten days when Christmas Island came over the horizon on July 11th. The next stop beckoned from 550 miles, the Keeling Cocos Islands, a small atoll in a limitless oceanic expanse. This was an acid test of navigation ability. Twenty-three days and 2,700 miles from Thursday Island in Torres Strait, Captain Joshua Slocum hit the Keeling Cocos right on the nose, arriving July 17th.

The next stepping stone was tiny Rodrigues Island, just west of Madagascar and 1,900 miles across the Indian Ocean. Captain Slocum sailed for the spot on August 22nd, and after the planned stop at Mauritius in the Seychelles plodded on to the southern coast of Madagascar. There the *Spray* locked horns with some of the heaviest weather since Cape Horn. She survived, as usual, although it was touch and go at times; and on November 17th she pulled into Port Natal. Word of the Slocum saga had reached here, too, and there was an enthusiastic reception by the Royal Natal Yacht Club.

The Long Way Home

One more great land mass waited to be rounded, the Cape of Good Hope. Here is an oceanic region off southern Africa where the Indian Ocean and South Atlantic collide, literally hurling themselves at each other. Its far-reaching "Cape rollers"—huge swells—are infamous, and it's an expanse noted for a surliness all its own. (I can testify to both rollers and moods, having fished the Indian Ocean off South Africa.) Captain Slocum understandably felt concern about what he might encounter. And on Christmas Day, 1897, his apprehension was justified as the *Spray* practically tried to stand on her bowsprit in Cape seas. Then, as if to atone for its rude behavior toward a distinguished visitor, the turbulence abated, and Captain Slocum dropped anchor in a welcomed calm at Cape Town. In the care of local authorities, the boat remained safely in drydock therefor three months while her owner toured the magnificent South African countryside via railroad.

Not until March 26, 1898 did *Spray* turn her stern to Cape Town.

After a brief recess at St. Helena, during which he spent a night in the room supposedly haunted by the ghost of Napoleon Bonaparte, Josh headed for lonely Ascension Island, dubbed "The Stone Frigate" by the British Navy. Ascension's sea-beaten rocks fell astern on May 8th, and the *Spray* at last was headed for home. On this leg occurred the voyage's climax, for it was then the 36-footer crossed her outbound track of October 2, 1895. She had circled the globe. You can imagine her skipper's elation.

They were still 3,500 miles from home mooring, however. Wind and currents seemed to aid the *Spray*'s eagerness to reach home port. They added a good forty miles a day, boosting her daily total to about 180 miles. Along the way came an interesting little meet that showed the skipper's sense of humor.

Communications being what they were in 1898, Captain Slocum hadn't yet heard about the outbreak of the Spanish-American War. Until he guessed the situation, it was a surprise when he first sighted the U.S. battleship *Oregon*, a great war machine of her era, steaming on her way to Santiago, Cuba, and combat. Turret guns already elevated to fire, the battlewagon ran up a Spanish flag and under it code signals asking Captain Slocum, "Have you seen any men-of-war?" *Spray* replied in the negative. As an afterthought, Joshua added code flags reading "Let's keep together for mutual protection." The *Oregon* was not sufficiently worried to take him up on this offer.

Forty-two days out of Cape Town, the *Spray* reached Grenada, then progressed to her last foreign port, St. John, Antigua. On June 14, 1898, just short of three years at sea, she cleared St. John for home and completion of her historic voyage.

Captain Joshua Slocum's fame had spread worldwide, and the next few years were profitable for him. He was encouraged to write a number of magazine pieces and syndicated newspaper articles. From his writing efforts also emerged a book that became a best-seller in its time, *Sailing Alone Around the World*. The skipper became something of a literary lion too, a status that was enhanced by lecture engagements.

Then a Blank

For what was to be her trip to oblivion the *Spray* was fitted out at the famed Herreshoff yards in Bristol, Rhode Island. It's important to mention that both skipper and boat were in the best of health when they sailed on that fateful autumn-winter visit to Grand Cayman. After circling the globe, this must have seemed like a hop-skip-and-a-jump to Captain Slocum.

And it was. Yet somewhere out there the valiant team of Slocum and *Spray* vanished for all time, leaving no clue to their fate. In light of all the extremely perilous situations they had survived, and the captain's enormous experience, you can see why the disappearance is in the nature of a mystery. The question is still asked: What *really* happened to the *Spray*?

Five possibilities present themselves: foundering in a storm; collision at night; fire; a freak situation in which Captain Slocum fell or was swept overboard, whereupon the *Spray* sailed on to a later death; and destruction by forces unknown.

The first alternative seems highly unlikely, because at that time there was no weather too severe for the skipper to cope with, and certainly not as bad as some he had encountered in his globe-girdling adventure. The third is always a possibility afloat. Yet this also sounds very unlikely, since the master mariner would have exercised every precaution with a stove and oil-fired lamps. No one could appreciate the perils of fire at sea more than a skipper of the sailing ship school. The fourth may have been more of a possibility than some observers realized. Normally it would be very unlikely that Captain Slocum would fall overboard or expose himself to the chance of being swept away. However, who can say that he didn't suffer a fainting spell or a heart attack, or slip on a wet deck and get knocked unconscious in a fall?

Writing about his father in 1950, Victor Slocum considered the second alternative, a collision at night, as the most likely cause. Collisions still occur, even with radar, and in 1909 there was no such electronic equipment. U.S. East Coast waters, notably those north of Cape Hatteras, have always been busy thoroughfares traveled by all kinds of vessels, fishing boats, freighters, and liners, with traffic lanes between Boston, New York, Philadelphia and Baltimore. In Captain Slocum's day a small craft's running lights were oil-fired—not too bright at best—and easily obscured by sails. In keeping with custom on sailing vessels, Captain Slocum always kept a turpentine-charged torch handy for quick ignition to reveal the *Spray*'s presence at night when necessary, but he may not have had a chance to use it. With failure of her lookouts to spot a small craft in the darkness, a coastwise steamer—perhaps poorly lighted herself—could have crept up on the *Spray* and run her down and demolished her without so much as a tremor to notify the larger vessel. Or the accident could have occurred in heavy fog.

The fifth possible cause, "destruction by forces unknown," I've saved until last because it adds a note of mystery and is timely. Barring a fatally severe local storm, it's difficult to imagine what those forces were. On the other hand, were Captain Joshua Slocum and his *Spray* victims of the Bermuda Triangle?

In Memoriam

In a sense, the *Spray* is memorialized by the mushrooming popularity of sailboating throughout America. It's to be wondered, though, how many amateur skippers would react to the name Joshua Slocum. If they were but a few it wouldn't be surprising. His era must seem like a thousand years ago to the majority of today's sailboaters. For later generations there was a reminder in the World War II liberty ship named in his honor; but even for many of those people who recognized his name the mists of time have all but cloaked memory of the man.

Still, courage and accomplishment are durable. Regardless of the failings of public memory, Captain Joshua Slocum will remain one of the greatest, and certainly one of the most daring, sailing skippers of all time—bar none.

26

Mystery of the "Oregon": A Hit-and-Run Sinking

Christopher Columbus and his bold company consumed some seventy days in their transatlantic voyage to the New World. Later mariners chipped away steadily at that crossing time, knocking off hours, days and, finally, weeks. By 1840, speed in traversing the Atlantic Ocean became a matter of maritime prestige when the Cunard firm's sidewheeler *Britannia* clocked a then-surprising eleven days four hours. Rightfully, Cunard made much of that accomplishment, and it was established as an international challenge. Also established was the Blue-Riband*, coveted recognition for the fastest crossings.

Cunard fractured its own record in 1863 when the *Scotia*, another sidewheeler, averaged fourteen knots to whittle still more time from the run. For the next sixteen years the Blue-Riband remained in British hands, thanks to the White Star Line's propeller-driven *Oceanic* and *Britannic*. They made sixteen knots the crossing speed to beat.

British monopoly on transatlantic speed honors finally was disrupted by an American named Stephen Guion. After thirty years experience with the Black Star Line in the United States and its Liverpool offices, and with Cunard, he went out on his own and founded the Guion Line. For enterprising Mr. Guion it was an obsession to bring the Blue-Riband to the United States. Accordingly, he planned ships that would be the fastest afloat.

Unfortunately, in his enthusiasm Stephen Guion was carried away to a bad start. He experimented with new and untried ideas, which was a worthwhile effort, except that the ideas were often unworkable too. Although his first two contenders, the *Dakota* and *Montana*, eventually found their way into the transatlantic competition, neither earned the Blue-Riband. In time, both ships ended their careers in watery graves, wrecked off the coast of Wales. But the Guion spirit remained undiminished, despite these setbacks and

*This is the correct spelling, not the more often used "Blue Ribbon." According to British maritime definition, "riband" signifies a ribbon awarded specifically for an accomplishment.

other heavy financial losses due to unwise engineering experiments.

Next to fly his line's flag were the *Arizona* and *Alaska*. The *Arizona* was first to appear, and her upcoming maiden voyage was given front-page publicity in English and American newspapers because of her unfortunate predecessors. There was snickering behind the scenes in the press; and some newspapers had humorous accounts ready, fully expecting this American upstart to reach Liverpool under sail, no threat to the Blue-Riband. At the time the White Star Line's *Britannic* ruled the waves, and no one looked for a ship from a young company to dethrone her.

As it worked out, newsmen—ship experts too—on both sides of the Atlantic had to eat crow. Much to the amazement of all knowledgeable observers—and the embarrassment of the more outspoken ones—the *Arizona* clocked the fastest Atlantic crossing to date, her westward run averaging seventeen knots. Stephen Guion's dream was realized. The Blue-Riband came to the United States.

The *Arizona* had further adventures. On one run she crashed into an iceberg at full speed, a meeting which pushed her nose back twenty-six feet. Even so, she remained basically seaworthy, and with a temporary wooden bow installed in Newfoundland she completed her round-trip crossing. During World War I she became the U.S.S. *Hancock*, ferrying U.S. Marines to Europe. Her colorful career ended in a San Francisco wrecking yard in 1926.

Soon after the *Arizona* laid her claim to the Blue-Riband, Stephen Guion launched her sister the *Alaska*. Fulfilling the line's expectations, this newest Guion ship cut transatlantic crossing time to less than seven days, averaging 17.76 knots. Despite a steady increase in speed, however, shipbuilders of that era just couldn't bear to do away with sails entirely. Consequently, all those sleek liners of the time were hampered in speed attempts by full-rigged masts and canvas which were used rarely, if at all.

Now the Oregon

Understandably, the Cunard people were disturbed by a much younger line taking away the Blue-Riband prize. It was much more than a matter of prestige. Passengers sought out the fastest ships, and transfer of their affections to a rival line cost Cunard many thousands of pounds sterling. The famed British shipping firm began work on two superliners. The *Eturia* and *Umbria* would be ocean-going hotels, fast and luxurious.

But before these giants of the era could be launched, the Guion Line sent its latest challenger, the *Oregon*, down the ways. Grossing nearly 7,500 tons, she was her owner's largest ship to date.

Moreover, the *Oregon* was looked upon as the bright hope—and maybe the last chance—to keep the Guion Line in business. Financial woes plagued the firm. Thanks to her speed, the *Arizona* maintained her line's solvency, but the margin between black ink and red was uncomfortably narrow.

The *Oregon*'s keel was laid down in 1881 at the yard of John Elders-Fairfield & Co. in Glasgow, Scotland. She celebrated her maiden voyage by capturing the Blue-Riband with an average speed of just under eighteen knots. She was the last American ship to win that coveted award until 1952, when the S.S. *United States*, superliner of the United States Lines, crossed the Atlantic in the record time of three days, ten hours and forty minutes, averaging an awesome 35.59 knots or roughly forty-one miles an hour. And that, in case you haven't guessed, is a lot of speed for 51,988 tons of "boat."

The *Oregon* was quite a lady in her day. She measured 540 feet long overall, with a 54-foot beam. Her engine could develop nearly 12,000 horsepower. It had three cylinders. One was a high-pressure chamber with a five-foot ten-inch diameter. The other two cylinders, each eight feet eight inches in diameter, operated on low pressure. The piston stroke was eight feet. Steam was generated by nine boilers, each eighteen feet long by sixteen feet in diameter, each heated by nine furnaces. Steam pressure tested to 220 pounds per square inch, but operated at approximately 110 p.s.i. This mass of machinery gulped coal at the rate of 242 tons a day to maintain a speed of eighteen knots. At the crest of her speed stardom the *Oregon* crossed the Atlantic in six days, nine hours and thirty-one minutes, averaging 18.16 knots. That was in August, 1884.

Although her two stacks were disproportionately tall, her profile bore a marked resemblance to liners of a much later date. Incongruous with more modern design, though, she stepped four masts, fully rigged for canvas. There still remained a reluctance to dispense with sails.

The *Oregon*'s speed, size and handsome appearance immediately captured the fancy of ocean travelers. Additional attractions were the posh refinements of her accommodations and public rooms. Her grand salon was an eye-opener in itself. Sixty-five feet long and fifty feet wide, it was a showplace featuring highly ornate woodwork fashioned entirely from timber cut in the state for which the ship was named.

Unhappily for her courageous owner, this epitome of transoceanic travel did not keep the Guion Line on the credit side of the ledger. By early 1884 the company was foundering in a sea of red ink. Creditors camped on its doorstep, and the line sank in bankruptcy. The *Oregon* was sold to the deceased line's rival, reportedly for 616,000 pounds or roughly $3,080,000. Cunard had bought out its greatest competitor in the Blue-Riband race. As it worked out, the award was to return to England anyway, won by Cunard's latest, the *Umbria*.

Ironically, the *Oregon* realized substantial profits for her new owner. Then, two years after purchase, she presented Cunard with its first major calamity in forty-three years.

Doomsday

The *Oregon*'s swan song came on March 14, 1886. The fatal incident began at 4:20 A.M., just before the first faint rays of sunrise.

She had sailed from Liverpool on March 6th. Aboard were 186 voyagers in first class, sixty-six passengers in second class, and 395 people traveling in steerage. Officers and crew numbered 205, with Captain Cottier commanding.

Passengers were still asleep during the early morning hours of the fateful day as the liner neared the offing of Long Island. That is, practically all the passengers were asleep. One, a Brooklyn man, had found sleep impossible and at 4:00 A.M. was taking a constitutional on deck. He had completed a lap and was standing on the afterdeck when he felt the ship shudder slightly underfoot, followed by what he described as a jolt that threw him off balance. When he heard shouting up forward, he hurried toward the bow section and came upon crewmen leaning out over the port rail and peering at something below. He too leaned out precariously over the rail and looked. Even in the poor light he could see a yawning hole in the ship's side "that you could drive a horse and wagon through," he said later.

The *Oregon* had been struck by another vessel.

By now a lady passenger had also become suspicious that all might not be well with the liner. Alone in Cabin 54, she was awakened by something—she didn't know what, she said in her testimony later—that startled her, and remembered seeing a "red flash" in her stateroom porthole at that instant. Next thing she knew, a steward was running along the passageway outside, banging on doors and shouting for everyone to hurry on deck immediately. Within minutes the passageway was feeding a stream of humanity toward the companionways topside.

Meanwhile, deck officer John Huston became a very busy man. Surveying the gaping wound in the ship's port side, he figured that the only procedure that could save the *Oregon* was immediate installation of collision mats to close it. No mats were at hand, however, so Huston attempted the next best thing, jury-rigging of canvas tarpaulins as a kind of "cork." He and second officer John Hood leaped into the cold ocean in a desperation effort to secure the tarpaulins in position. At first they succeeded, but under pressure the canvas didn't hold properly, and water continued to pour into the hull at an alarming rate.

While officers Huston and Hood were struggling with their emergency contrivance and passengers scurried topside, a severe complication erupted in the ship's bowels. The liner's wound opened into her largest single compartment. This was located below the dining salon on the port side, just abaft the foremast, and measured 127 feet long. Ocean water rushed into this large compartment, drowning fires under boilers and generating clouds of steam amid a black fog of coal dust. Men in the engine-room gang panicked, which wasn't surprising considering that water was climbing to knee level and live steam scalded their faces. They bolted for ladders leading topside out of that watery-steamy nightmare, fighting and clawing at one another to clutch the rungs. Some later testimony indicated that they reached the deck before any women and children appeared, since it was said that they scrambled to the nearest lifeboats, launched them, and rowed away like mad.

Now the *Oregon* was exhibiting a slight list.

Fortunately, calm heads prevailed on the bridge. Captain Cottier had been in his quarters at the time of the collision, leaving his chief officer, a man named Mathews, in charge. Weather was clear and the sea calm. Later Mathews was to state that a coastal schooner, running without lights, appeared suddenly out of the gloom and slammed into the *Oregon*. He summoned his skipper immediately and ordered rockets and guns fired as distress signals. Two steamers could be made out on the horizon, but it was reported that neither altered course in response to the liner's signals. At the time of impact the stricken *Oregon* was only about five miles off Fire Island, New York. Captain Cottier ordered a change of course, intending to run the ship aground on the island. But power was lost very quickly when water drowned the boilers' fires, so this emergency procedure was impossible. At this point arises an unsolved riddle: Why didn't the *Oregon* use her sails to reach shore?

In contrast with the engine room gang's frantic exodus, the passengers remained cool, calm, and collected. Many were even in good spirits. Most of their luggage already lay on deck in anticipation of docking in New York later that day. Those who were being light-hearted obviously weren't aware of their predicament. They got the message when officers apprised them of the situation and urged that they obtain warm clothing from their staterooms. For many it was too late to go to their cabins. Securing of watertight doors had sealed passageways.

"Abandon Ship!"

Operations were begun to evacuate the *Oregon*. After the engine-room force's hasty departure, eight large lifeboats remained. The air was cold, thirty-five degrees, but not biting. There was one small measure of cheer: The sun was coming up; it soon would be daylight and a bit warmer. "Women and children first!" came the order for abandonment.

Pretty soon there was a large measure of cheer. Substantial aid had arrived on the scene. Standing by was *Pilot Boat No. 11*, which had come out looking for the *Oregon* to guide her through Ambrose Channel and the approaches to New York Harbor. Joining the pilot vessel was the schooner *Fannie A. Gorham* out of Belfast, Maine, a Captain Mahoney commanding, en route from Jacksonville, Florida, to Boston. Although the liner now listed much more noticeably, transfer of passengers to those two vessels was accomplished without mishap. Practically without mishap, that is. Again deck officer Huston entered the picture, this time as hero of the transfer operation. He was credited with saving at least three lives: two children who lost their footing and dropped into the sea, and an elderly man who fell overboard from a lifeboat.

The *Oregon* was on her way to a grave, inching lower and lower in the sea. *Pilot Boat No. 11* and the *Fannie A. Gorham* were still receiving passengers when the North German Lloyd Line's steamship *Fulda* nosed over the horizon and hurried to the scene. Her appearance was providential, because by now the pilot boat was dangerously overloaded. A possible disaster was averted when the *Fulda* took a number of people off the overcrowded vessel.

Although an *Oregon* seaman later stated that seven of his shipmates went down with the ill-fated liner, official records indicate successful rescue of all on board. True to the code of the sea, Captain Cottier was the last to leave the doomed ship. It was very considerate of the *Oregon* that she dallied for eight hours before slipping beneath the waves forever. Down in about eighty feet of water, three of her masts marked her death site for days afterward.

Two Mysteries

The *Oregon*'s most precious cargo, passengers and crew, was saved from King Neptune's clutches. But consigned to the bottom with her went a shipment of machinery, hardware, silk and other drygoods, liquor and costume jewelry appraised in excess of

$700,000. In addition, she carried some 757 mail sacks containing an estimated one million dollars in paper currency and twice that amount in stocks and bonds. Rescued by the purser and crewmen was a valuable consignment of diamonds reposing in a ship's safe. About seventy-five mail sacks were retrieved before the *Oregon* sank. Many more were picked up by vessels afterward, but their contents already were badly damaged by submersion.

According to two prominent Long Island divers and undersea historians, Charles Dunn and Graham Snediker, in an article submitted to me several years ago, the Cunard Line listed a vessel named *Charles R. Moss* as rammer of the *Oregon*. So far as I can determine, though, the culprit's identity was never proved and remains a mystery. Described as a schooner, the mysterious adversary sank very quickly after tangling with the liner's steel hull. At a hearing, *Oregon* crewmen testified that they saw a sizable schooner smash into their ship and sink within three minutes, but not before they heard screams from men aboard her.

The ramming schooner's name was one enigma. Another was that no wreckage or other debris, or any survivors or bodies, were found. Nothing, that is, which could be traced definitely. On March 21st a schooner's yawl was discovered, derelict and adrift, about twenty-five miles from the sinking. Lines still attached looked as though they had been severed with an ax, hinting that the yawl left her parent vessel in a big hurry. The craft carried no name or identification. Some flotsam from the *Oregon* also was picked up in the same general area, but none from the mystery vessel.

Repercussions from the accident bounced around for months afterward. All sorts of accusations, charges and countercharges flew through the air. Some passengers tried to place the blame on Cunard. Others were bitterly critical because no attempt was made to beach the liner. There were confusing contradictions. One passenger testified that he saw the lights of an oncoming vessel for an hour before the collision. Some of the *Oregon*'s crew, you'll recall, stated that the other vessel carried no lights. Accusations went so far as to claim there hadn't been a collision at all, but that the *Oregon* was rent by an internal explosion—this despite evidence to the contrary. Some charges bordered on the ridiculous.

In time an official inquiry convened in Liverpool. No passengers were permitted to testify, but masses of evidence were sifted and evaluated by a panel of maritime experts. Their findings created the one known human casualty in the *Oregon*'s sinking.

The board of inquiry cleared other officers of blame for the collision, but found reasons to zero in on Captain Cottier. Only forty-five at the time, he was the youngest master in Cunard service.

But he had long years of experience, having held a master's papers for twenty years. What's more, he was rated as an excellent commander and navigator, and had been with Cunard seven years. All this cut no ice, however. Cunard held him responsible, and he was sacked.

Eventually a salvage company dispatched tugs to the *Oregon*'s area to probe chances of raising her. Divers descended to the wreck and examined it, then reported that the ship already had broken into halves, eliminating any possibility of retrieving her.

And there the *Oregon*, former speed queen of the Atlantic, has rested and disintegrated ever since. For many years she was a fishing ground for anglers above the waves, and an attraction to skin divers below.

27

Deadly Secret of the "Morro Castle"

On a September morning in 1934 citizens of Asbury Park, New Jersey, received two surprises, one pleasant, the other very unpleasant. The nice surprise was sunshine after a storm that howled at hurricane intensity at times. The nasty surprise was a close-inshore grounding of the still-smoldering corpse of a once majestic ocean liner, the *Morro Castle*.

Word of the disaster had flashed far and wide much earlier; but only the most inquisitive and hardier residents of the resort city braved the wild wind and drenching rain of the previous night to watch the ravaged hulk, illuminated by flames, drift shoreward and threaten to demolish the town's convention hall, which extended out over the water. With passage of the storm, crowds quickly gathered on the beach to ogle the dead liner, now firmly stranded. Thousands more converged on Asbury Park via every conceivable route, creating monumental traffic jams.

What they beheld was the pitiful, smoking result of one of U.S. maritime history's major calamities. What they couldn't see, and what was never revealed with any certainty even to expert investigators, was how the catastrophic fire started.

Turn Back the Calendar

In 1928 the United States was still riding high, wide, and handsome on national prosperity. This was an era of freewheeling spending. Indeed, it was one of the boom times in the nation's history.

Behind this prosperity's facade, however, the U.S. Merchant Marine was suffering abject poverty. Its giant fleet of vessels that mushroomed during World War I had dwindled to a flotilla: A few passenger liners, not many more freighters. United States lines were losing freight and passenger business right and left to the lower rates and better service of foreign-flag ships. In Washington there was mounting concern, not only about the economic aspect of enormous revenue losses but also about the cost in U.S. prestige. Added was a military concern. In contrast to the merchant fleet, the U.S. Navy was of such size and power that it challenged England's age-old boast, "Britannia rules the waves!" But responsible observers, harking back to the latest war, realized that a navy must be backed by a suitably large merchant marine.

In the spring of 1928 Washington responded by enacting the Merchant Marine Act, which provided a then-huge pool fund from which private shipping companies could borrow for construction of ships. Under this act's terms, every ship's plans had to be approved by the Navy before construction began. Further, it was required that passenger liners be designed for ready conversion to troop transports in event of war. Along with the availability of loans for financing, the federal government endeavored to encourage expansion of the merchant marine by increasing the subsidies granted to vessels carrying U.S. mail.

Of this revitalization was born the S.S. *Morro Castle*. Her owner, the Ward Line, known more formally as the New York and Cuba Mail Steamship Company, had been toting cargos, passengers, and mail between New York and Havana for a hundred years, but was sadly in need of new ships. Its *Mohawk* and *Orizaba* was elderly and slow. On an average they consumed three days for the New York-Havana run, which figured out to about sixteen miles an hour. In a way, this slow rate was fine for passengers. In those years the United States was in the grip of a fierce thirst—Prohibition; and with liquor available aboard ship, the longer a cruise, the more time to slake that thirst. For the Ward Line the slower running time was bad news. The *Mohawk* and *Orizaba* were like cows on a race track where horses had been introduced. The company began to lose patronage to faster foreign-flag liners such as the *Mauretania* and *Europa*.

Master naval architect Theodore Ferris, whose firm had designed many ships of different kinds, was commissioned to execute plans for two Ward liners: *Morro Castle* and a sister, *Oriente*, named for the fortress guarding Havana's harbor and one of Cuba's provinces. (Relations between the U.S. and Castro's land were happier in those days.) Ferris was directed to spare no expense in designing liners that would be the latest in safety, speed, passenger accommodations, and freight handling. The sister liners would be 508 feet long, with a 70-foot breadth, and gross 11,500 tons apiece. Each would accommodate 489 passengers and be handled by a crew of 240 officers and men. And they would be the fastest vessels of their type on the U.S. Atlantic Coast.

Designer Ferris came through on all counts. Not only would the sister ships meet U.S. Navy requisites for safety and dependability, they would also slice at least half a day from the New York-Havana run while transporting their guests in luxury. Although they would be relatively close to the coast in their shuttling back and forth, Ferris wanted to accent safety. Features included watertight compartments with doors that slammed shut automatically. An elaborate fire protection system incorporated automatic foam extinguishers in holds and engine-rooms, tubes to carry any warning of smoke from cargo compartments to the liner's control center, and an electrical sensor system that would detect fire in any of the ship's 217 staterooms as well as in officers' and crew's quarters, and flash warning signals on a monitored panel. And there were check points throughout the vessel at which watchmen would regularly ''punch'' their clocks.

To fight fire, the *Morro Castle* and her sister each were outfitted with numerous hydrants at intervals throughout five decks, literally miles of hose, and upwards of 100 extinguishers. At strategic locations were bulkhead doors that closed automatically to contain a fire.

Just in case it ever became necessary to abandon ship, a possibility that seemed remote, architect Ferris wanted the ships to be well prepared. Each liner would carry twelve lifeboats, some motorized, with a collective capacity of 826, lowered by simply throwing a lever; an equal number of roomy liferafts; numerous lifesaving buoys of the standard ring type; and approximately 850 life preservers, 121 more than required by passengers and crew put together. In all, it could be figured that the *Morro Castle* carried lifesaving gear for some 1,900 people.

Yet an uncontrollable fire *did* break out; there *did* come a time to abandon ship; and there *was* a shocking loss of life. Root of the tragedy, it appears, lay in a fatal omission: No fire-detecting sensors had been planned for some of the ship's public rooms, including a lounge, dining room, library, and ballroom. The start of the disastrous fire was traced to one of these, a locker in the writing room.

Success with a Bleak Background

Theodore Ferris's plans promised two of the finest, fastest, safest luxury liners afloat, and in midautumn of 1928 the United States Shipping Board approved loans for construction of the *Morro Castle* and *Oriente*. Keel of the *Morro Castle* was the first to be laid down. Her steel skeleton took form starting in January of 1929 at a yard in Newport News, Virginia, one of the East Coast's great shipbuilding centers.

The *Morro Castle*'s outer form was taking shape and her turboelectric engines were being installed when there came a financial crash heard around the world. On October 29th, the terrible Black

Thursday, 1929, the New York Stock Market collapsed under the sheer weight of unrestricted buying on margin. In one day thousands of investors were wiped out; many thousands more followed. It was said for a while that to walk along Wall Street was to risk being hit from above by a fast-falling body—that of a broker or investor.

So was born the grim depression of the 1930s. One would think that Black Thursday and its aftermath would bode ill for a luxury liner still being built, but the nationwide economic calamity didn't halt construction of the *Morro Castle*. She was launched in 1930 amid lots of fanfare and publicity. Fortunately for her beginning, the depression had not yet tightened its vise fully.

On a shakedown trial in early August 1930, the *Morro Castle* realized a speed of nearly twenty-one knots, meeting her designer's expectations. All else being found satisfactory, she was placed on exhibition in New York, and on August 23rd she was ready to embark on her maiden voyage. On her initial round trip the new superliner pleased eveyrone by knocking about fourteen hours off the run in each direction. With her sister *Oriente*, launched in May of 1930, the *Morro Castle* entered into fast, luxurious cruise service to and from Cuba; and with elaborate menus and bars which stayed open until all hours, dispensing whiskey at twenty-five cents a shot, they became very popular ships. What's more, the *Morro Castle* proved herself very seaworthy, surviving a hurricane that practically followed her from the Florida Keys to the offing of Cape Hatteras.

Sunshine and Shadows

Command of the *Morro Castle* was given to Captain Robert R. Willmott, a veteran of twenty-six years with the Ward Line. His reputation was that of a very capable skipper, respected by his profession. The crew, however, seems to have been a question mark—at least in regard to some of its members. Recruitment could be rather haphazard in those days. Enforcement of requirements for seamen was lax. It was said that it was even possible to buy a certificate stating that its holder was qualified. Further, there was nothing in seamen's documents to indicate whether or not they had criminal records. Still further, seamen's pay then was poor, and reportedly those aboard the *Morro Castle* sometimes complained of unsatisfactory chow. It's to be wondered whether or not some of those details were significant in light of subsequent events.

The fateful final cruise was a return from Havana in 1934. Although the Great Depression was at its height, with millions of workers jobless, men peddling apples for five cents each on street corners, and the song "Brother, Can You Spare a Dime?" practically a national anthem, there were still people with the wherewithal for a pleasure cruise to Cuba. When the *Morro Castle* glided out of Havana harbor on September 5th for her date with destiny she carried 316 passengers, served by a crew of 232.

At this point a weird note insinuates itself. Captain Wilmott, normally gregarious and fond of food and conversation, gradually became noticeably irritable and inward-drawn. He seemed apprehensive about something, and grew increasingly suspicious of certain officers, notably Chief Radio Operator George Rogers and an assistant. On the morning of September 7th he privately voiced vague suspicions of possible sabotage of the liner to First Officer William F. Warms and Fourth Officer Howard H. Hansen. The "old man" appeared to be badly upset, Hansen was to recall later.

September 7th would be the last night at sea on that cruise, traditionally an evening of festivities, a last fling before docking. An indication of Captain Wilmott's emotional or physical upset came when he deliberately avoided the captain's dinner, a last-night-at-sea affair, and picked at his meal in the seclusion of his quarters. Not long after a steward delivered his food, Captain Wilmott spoke by ship's telephone with the *Morro Castle*'s surgeon, Dr. Van Zile, and complained of "a little stomach trouble." That turned out to be an ominous remark. Just before 8:00 P.M. Captain Wilmott was found dead in his cabin, the apparent victim of a heart attack. Chief officer Warms automatically became the ship's acting commander. Word of the skipper's death was announced to the passengers, and the evening's festivities were canceled. Most of the passengers then retired to their staterooms earlier than usual, a circumstance that subsequently contributed to catastrophe.

Meanwhile, the elements were conspiring to create a suitable backdrop. On September 6th, roughly twenty-four hours out of Havana, ominous-looking clouds started to gather on the southern horizon. By the following morning they grew heavier and moved in. The wind shifted from south to east; a light rain spattered the liner. Her weather-wise officers recognized the symptoms of a brewing storm, a bad one. And as the hours passed their diagnosis was confirmed. The storm intensified, sending a number of pale- or chartreuse-faced passengers to their beds with seasickness. By dinner time that evening the *Morro Castle* was in the grip of a nor'easter; and on the U.S. Atlantic Coast a nor'easter can be almost as rough as a hurricane.

It was hardly an occasion for last-night-at-sea festivities. Captain Wilmott was dead, his beloved ship was being clobbered by a worsening storm, and commanding her was an officer who had
207

never been in complete charge of this liner before. Acting skipper Warms' personal concern was understandable. Suddenly he was in charge of a multimillion-dollar vessel with 548 humans aboard. But the storm itelf created no great worry yet, because the *Morro Castle* had come through worse before.

Fire!

In the writing room just off the lounge on the port side of B Deck was a large locker in which a gross or more of newly cleaned blankets were stored pending colder weather. It isn't clear who smelled smoke first, but at 2:15 A.M. a passenger looking for a friend in a nearby bar detected it and set out with a steward to trace its source. They found it in the locker containing the blankets. Smoke billowed from around its doors, and the interior was already a mass of flames. The passenger triggered a fire alarm while a steward tried to put out the flames with a hand extinguisher. Minutes later, crewmen rushed in with a fire hose. According to witnesses, there was no pressure from the hydrant, and the hose issued only a trickle of water.

Very quickly the locker fire was out of control and spreading through the ship's midsection like a prairie blaze fanned by a Dakotas wind. Flames fed on the ornate wooden paneling that added so much to the liner's decor. Ravenously they consumed the studding between panels and metal bulkheads, racing through those air spaces. Layers of paint on superstructure and hull added more fuel. Soon the *Morro Castle*'s entire midsection was a furnace. Metal bulkheads became superheated and spread the conflagration to other parts of the now doomed liner. The automatic fire doors failed to contain the flames. Investigators theorized that these may have been on manual at the time, for some reason, or that they might not have prevented the spread anyway, because of the superheated bulkheads.

Describing the scene that ensued as a nightmare is an understatement of the century. Picture, if you can, a ship that became a furnace in the middle of a severe storm. On the fire-control system's panel, light after light flashed, indicating ignition of passengers' cabins. Acrid, choking smoke filled corridors and billowed into stairways between decks. Metal bulkheads glowed with heat. Decks became unbearably hot underfoot. Even the storm sky was lighted orange red by flames from the *Morro Castle*. Calamities piled on top of calamities. The nor'easter's wind was like Satan's bellows, fanning the fire into a holocaust. Soon the gyrocompass, electric steering system, and standby hydraulic system were burned out, and telephone communications were rapidly on their way to nonexistence. Deep in the ship, far aft between the propeller shafts, was emergency steering apparatus that required six men to work. It still could be reached; but, whether there were doubts about its accessibility or communications failed, no order for its operation came from Captain Warms. Shortly after 3:00 A.M. he adopted the only procedure he thought he had left, steering the ship with her turbines. Distress blasts of her whistle sounded above the wind's howl.

Horrors Multiply

At various distances ahead of the *Morro Castle*, and astern, traveling northward, were the Furness Line's sleek *Monarch of Bermuda*, the new 600-foot freighter *Andrea Luckenbach*, and a smaller ship, *City of Savannah*. They picked up the doomed liner's SOS and made ready to come to her assistance. In time the Dollar Line's cruise ship *President Cleveland* also moved into the scene. Meanwhile, the *Morro Castle* had shut down her engines and anchored in preparation for abandoning ship.

It's impossible to fully imagine the scene aboard the *Morro Castle*. Towers of flame leaped skyward from a fire still unsated. Billowing smoke mixed with rain as the storm continued in its fury, churning the ocean into fifteen and twenty-foot waves. Passengers' emotions ranged through a full spectrum, from amazing calmness to screeching terror. Plunged into a situation with which very few skippers have to cope, Captain Warms was assaulted by hideous problems from every quarter. According to some accounts and testimony, a couple of his key officers failed him out of panic; and the crew was split by fear, confusion, and a desire to help but not knowing how. Pandemonium was in charge of the *Morro Castle*.

Her engines shut down, her wireless silent, and her vital electrical systems destroyed, the floating furnace was helpless when the order came to abandon ship. Many passengers already had done so, preferring to take their chances in the sea's maelstrom than be roasted alive. The fire already had claimed victims; so had smoke inhalation. Contributing to confusion, terror, panic, and the toll of deaths and injuries was the fact that many passengers had been roused from deep sleep. Still fuzzy in their thinking, they bumbled into calamity or were overtaken by it before their heads cleared. Ironically, a few of those who jumped overboard were killed by the life preservers intended to save them. In the form of a vest, these preservers contained large blocks of cork. If not held down when leaping from a height, they could be rammed up under the wearer's chin on impact with the water, breaking the wearer's neck or knocking him unconscious, causing him to drown.

Three of the liner's lifeboats jammed in their davits and couldn't be launched. A couple of those that were launched, it was charged later, had plenty of room left but moved away without picking up survivors struggling in the water. Drowning claimed victims; so did submersion and exposure. Conceivably, some died in the water from injuries, shock or heart failure, or drowned when overcome by fatigue. Survivors told about seeing people caught in portholes and burned alive as they attempted to escape. It's to be wondered if any passengers were sucked into the ship's propellers by leaping from the stern while the *Morro Castle* was still under way.

A number of vessels eventually gathered at the disaster scene. They included the ships previously mentioned, along with the Coast Guard cutters *Cahoone* and *Tampa* and a large New Jersey sport fishing boat, the *Paramount*. They plucked living and dead from the water, but rescue attempts were seriously hampered by the storm. Amazingly, some of the hardier—and luckier—survivors managed to reach shore under their own power. And, as usual in catastrophes, there were stories of courage, heroism, and cowardice.

The late Captain Wilmott unintentionally fulfilled a prophecy he once made at dinner aboard. A guest at his table, sensing how proud Wilmott was of the *Morro Castle*, asked him what he would do if he ever had to leave her. The skipper laughingly replied that he would take her with him. And so he did, in a manner of speaking.

Tab for a Tragedy

There are variations in the reckoning of the number of lives lost in the *Morro Castle* disaster. A Coast Guard report's figures mention 86 passengers and 49 crew members. A plaque beneath the liner's bell, displayed in the New York offices of the American Institute of Marine Underwriters, sets the toll at 94 passengers and 30 crewmen.

The *Morro Castle*, of course, was a total loss, a multimillion-dollar loss. In addition to nearly five million dollars insurance underwriters had to pay out on her, claims for cargo and other losses exceeded $1,500,000.

An earlier attempt to tow the *Morro Castle* into port ended in failure when the tow cable snapped under the hulk's dead weight and the storm's force. Already a total loss, she burned for days after drifting ashore at Asbury Park. Even in death her evil star continued to shine. During preparations to get her out of there in November, an officer in charge fell dead from a heart attack and his body plunged from an upper deck.

In mid-March, 1935, tugs finally dragged the huge, charred corpse from its sandy cradle to a temporary anchorage in Brooklyn.

The possibility of rebuilding was considered, then rejected as being financially unfeasible. The pathetic remains of the *Morro Castle* were sold as so much junk, towed to Baltimore, and demolished for scrap metal. Still the liner's jinx wouldn't abandon ship. During demolition in a Baltimore yard the hulk caught fire again, this time from a combination of a blowtorch and oil in the bilges. Extinguishing it took some doing, but there was no loss of life.

Aftermath

Investigation of the disaster began that autumn and was intensive, complex, and long, involving the United States Attorney General's office and the FBI, the latter entering the case to find and interview survivors. Hearing after hearing produced a mountain of testimony from the liner's officers, crew and passengers. Some fifty-seven witnesses were questioned.

When it was all over, five *Morro Castle* officers were suspended, at least temporarily. The court leaned on three of these men, handing down terms in federal prison to two and fining the third five thousand dollars. In April 1937, the U.S. Court of Appeals reversed the prison convictions of the two officers, ruling that they had done everything required of them in the emergency. The five-thousand-dollar fine of the other officer was upheld. Acting Captain William F. Warms returned to the sea as skipper of a Ward Line freighter. One might wonder if misfortune followed him from the *Morro Castle*. His ship ran aground on a coast in Mexico, and he died in 1953. One of the officers whose conviction was reversed never got another chance. A few years later he passed away in retirement and obscurity.

Things didn't go too well for the *Morro Castle*'s owner either. The line was fined ten thousand dollars, maximum under the law, and eventually was dissolved.

It was never established with certainty what really caused the fatal fire on the *Morro Castle*. A plausible theory is spontaneous combustion of cleaning fluid among blankets stored in the writing room locker. Malfunctioning of an overheated funnel was thought a possibility. Defective electrical wiring had to come in for consideration. Arson was mentioned, suggesting a disgruntled or demented crewman as the villain. Yet another theory proposed sabotage, with a delayed-action incendiary device as the igniter. Some of this theory's subscribers cited a possible link with trouble in Cuba, since the island was in the throes of a revolution at the time. You'll recall that Captain Wilmott also entertained suspicions of possible sabotage, although his focused on certain crew members. Nothing concrete was found to support this theory.

They were driven home the hardest way possible, but the *Morro Castle* disaster did teach lessons of vital benefit to future sea travelers. It taught valuable lessons in prevention and control of fire aboard ship. It brought refinements in lifesaving equipment and procedures. It spawned more stringent laws to assure safety at sea and guarantee qualified personnel to man ships. Most important, it brought about a determination that there should never be another *Morro Castle* disaster.

Ripples

Throughout the *Morro Castle* story runs a bizarre thread in the form of her chief radio operator, George W. Rogers. In accounts he appears as a kind of shadowy character, a contradictory figure, a loner not liked especially by his fellow crewmen. Moreover, Captain Wilmott seems to have viewed him with suspicion, although it is not clear why. It may or may not be significant that one of Rogers' hobbies was the study of explosives and incendiary devices.

A large man in both height and girth, Rogers emerged from the *Morro Castle* disaster a hero, and was locally acclaimed as such. Elsewhere, there were people who had reasons to doubt this status.

Even those who acclaimed his heroism must have been jolted by these events in his later career. A strange fire gutted his shop when he set himself up in business as a radio repairman. After appointment to a local police department as a patrolman, his superior (Rogers was in line to succeed him) received an explosive device in the mail that crippled and almost killed him. In connection with this, Rogers' known interest in explosives and other evidence sent him to New Jersey State Prison. After just under four years' incarceration, he walked out of the penitentiary. During the summer of 1953 he reportedly began flashing a lot of money around his New Jersey town. Subsequent detective work found two of his close friends, a well-to-do elderly man and his spinster daughter, bludgeoned to death in their home, a few doors away from Rogers. The erstwhile "hero" of the *Morro Castle* was convicted of first-degree murder and returned to New Jersey State Prison with a life sentence. There, in January of 1958, the corpulent Rogers died from natural causes.

In fairness, certain details should be mentioned. First, Rogers was not seen near the locker where the fire was discovered. And it would have been difficult for a man of his bulk—he was six foot two and close to 300 pounds—to sneak an incendiary device into the locker between the watchman's half-hourly visits on his rounds without being seen by passengers wandering around the nearby lounge and other public rooms. It was possible, but improbable. He had no apparent reason for such an act, at least none that was known. Arsonists are noted for hanging around to observe the results of their handiwork. Rogers remained at his post in the ship's radio shack. If Rogers set the fire, he soon had reason to believe that he too might become one of its victims. Why did he remain at his transmitter instead of being one of the first to leave the ship?

Despite those details, the detective crippled by the mailed explosive device and a writer collecting material for a book about the *Morro Castle* couldn't separate Rogers from guilt for the fire. There was his keen interest in explosive and incendiary devices, and Rogers once described in detail to the detective how an incendiary "pencil" could be made. Rogers was obsessed by the disaster. In light of the fact that he was something of a braggart, it may have been significant that he steadfastly maintained that the fire was started by an incendiary "pen" in the pocket of a jacket hung in the writing room locker.

The writer interviewed Rogers in the New Jersey State Prison, Trenton, and asked him point-blank if he started the *Morro Castle* fire. He didn't admit it, but he didn't deny it either.

If the *Morro Castle*'s deadly secret was his guilt, George Rogers took it with him in death within the bleak confines of a penitentiary.

28

What *Did* Happen to Amelia Earhart?

If the modern Women's Liberation movement had been in existence in the 1920s and 1930s it would have derived considerable pride, nourishment and ammunition from famed flyer Amelia Earhart. Time hasn't discounted the value of that lady to her sex, however. Amelia Earhart's accomplishments in aviation will endure for as long as there are women and aircraft.

Some Twists, Then Wings to Fame and Glory

Amelia Mary Earhart was born in Atchison, Kansas, on July 24, 1898. When she graduated from a Chicago high school in 1916 she already possessed a drive for personal accomplishment and a resolute determination be completely self-reliant. Those two motivating forces were to become as much a part of her as her all-American face.

The year after graduating from high school she went to Canada as a volunteer aide in tending World War I wounded in a Toronto military hospital. This was the first major twist in her path, for it steered her toward a career in medicine. In 1919 she enrolled at Columbia University in New York to begin premedical studies. If fate hadn't intervened, she would have become Amelia M. Earhart, M.D., because she had a habit of attaining her objectives.

Destiny tapped her on the shoulder while she happened to be in Los Angeles, where her parents had moved in the meantime. Quite by chance she attended an air show, which in those days featured daring aircraft races and aerial acrobatics. Now Amelia was gripped by a new and more powerful interest that pushed aside a career in medicine. And when she had a flight in one of the air show's planes, she landed with a route to her future firmly fixed in her mind.

Amelia took flying lessons from a pioneer aviatrix with the unusual name of Neta Snook, first female graduate of the Curtiss School of Aviation. From there she went on to advanced training with a former U.S. Army flight instructor, and in 1921 made her first solo hop. The next year brought another milestone. With her dad's financial help she purchased the small biplane that launched intense pursuit of a career as a professional woman flyer. Also that year came the first of the crashes from which she was to walk away, not the least bit deterred or discouraged. In fact, only a few days after her first crash she zoomed aloft to capture her first record, climbing to 14,000 feet to establish a new women's altitude mark.

This was to be the story of Amelia Earhart's life: An insatiable love of flying and a continuing series of notable accomplishments.

Aviation "fever" swept the United States in the 1920s. Cmdr. Cummings Read and a crew of five had fired the public's enthusiasm with a transatlantic crossing in the U.S. Navy "flying boat" NC-4 in 1919. Aviation "firsts" followed in awesome profusion during the 1920s, climaxed by the heroic solo venture of Charles A. Lindbergh, the "Lone Eagle," from Roosevelt Field on Long Island, to Paris in May of 1927. A new breed of heroes was born, and the public went wild in its acclaim.

Amelia Earhart stepped—or flew—into this milieu with ease. Already a seasoned pilot, she immediately had a place in what traditionally was an all-male field. What's more, she earned the respect and admiration of her male cohorts. And she remained a feminine woman, her profession, her close-bobbed hair and her aviation attire notwithstanding.

On June 17, 1928, she became the first woman to cross the Atlantic Ocean by air from Newfoundland to Wales. She was keenly disappointed that she had to make this flight as a passenger, not as the pilot, but it earned her instant and overwhelming notice. In a matter of hours she became one of the world's most famous women.

From then on her career was one headline-proclaimed feat after another. In May of 1932 she aimed her red Lockheed Vega at the sky over Newfoundland and in just under fifteen hours, after hair-raising threats from foul weather, ice on the wings, and possible engine trouble, touched down in a field outside Londonderry, Ireland. Amelia Earhart was the first woman to fly alone across the Atlantic.

Her fame mushroomed, bringing audiences with royalty, dinner with President and Mrs. Herbert Hoover, and an avalanche of medals, citations and other honors. She received her country's Distinguished Flying Cross, the first woman in U.S. history so honored. Even before the tumult of public adoration subsided, she set a women's nonstop transcontinental speed record: Los Angeles to Newark, New Jersey, in nineteen hours and five minutes. The following year she did it again, clipping almost two hours off her own record. She won still more women's solo-flight honors in 1935.

215

So did a pattern of triumphs pave the way for what would be her supreme accomplishment: A flight around the world.

So too was the stage set for what was to become one of the most fascinating sea mysteries of modern times: the disappearance of Amelia Mary Earhart.

Preparations

The year was 1936. By then, "A.E.," as she was known to intimates, was married to book publisher George Palmer Putnam. The proposed around-the-world flight received his endorsement and financial support.

For the great adventure Amelia selected what was then considered the most advanced long-range aircraft outside the military. Of low-wing monoplane design, it was a twin-engine, dual-control Lockheed 10-E Electra. It was rated to carry ten passengers, giving ample room for a navigation compartment and extra fuel tanks. The original purchase order, dated March 10, 1936, specified a total fuel capacity of 1,204 gallons, including wing tanks, for a maximum cruising range in vicinity of 4,500 miles. A.E. took possession of the Electra in July.

Preparations were complex and lengthy. Besides readying the plane, arrangements had to be made for fuel and parts depots at various stages of the flight. Clearances from the U.S. Department of State were required to visit various countries. Charts for the entire flight (it was decided to circle the globe from east to west) had to be studied and plotted with courses.

Two crack navigators were lined up. One was Captain Harry Manning, skipper of an ocean liner, whom Amelia had met while returning from Europe. He had volunteered his services on her future flights when on leave from his maritime duties. The other was Cmdr. Fred J. Noonan, rated as a genius in aerial navigation. It was planned that these two men would accompany Amelia at least as far as Australia. At the last minute Paul Mantz, an ace pilot in his own right and the world flight's technical advisor, volunteered to go along as copilot on the first leg, from Oakland, California to Honolulu. He wanted to join his fiancée, vacationing in Hawaii.

On St. Patrick's Day, 1937, everything was set. After a delay by weather until late afternoon, the Electra emerged from its hangar and everyone got aboard. With Amelia at the controls, the Lockheed's powerful Pratt & Whitney engines roared, and aircraft No. NR 16020 leaped forward eagerly. Its 55-foot wingspan providing lift, the mechanical bird banked gracefully toward the open Pacific and Hawaii.

Changes in Plans

The flight to Honolulu was uneventful and consumed just under sixteen hours. While the others rested briefly, Mantz shifted the Electra to Luke Field, which had a longer runway. The plane's next destination was tiny Howland Island, only two miles long by a half-mile wide, not even a flyspeck in the Pacific's vastness. Howland lay far out across the open sea. The navigational accuracy and hazards involved were awesome. Additional fuel had to be taken aboard the Electra as a precaution, and that necessitated a longer runway to become airborne.

Takeoff began as scheduled in the dawn hours of March 20th. Its first minutes were its last. The Electra didn't gain sufficient speed, and quickly was in trouble. Despite expert efforts to correct it, the plane began to lurch wildly, then slewed in a wide circle, coming to a grinding, crunching, shuddering halt. Landing gear, one wing and propellers were damaged, and fuel poured from ruptured wing tanks. Fortunately, Amelia had cut all switches, preventing what could have been a disastrous fire. Once again she walked away from a crash. Beyond a shaking up, no one was injured. Observers variously blamed a blown tire, improper distribution of additional fuel, and pilot's loss of control. An official report simply said "unexplained circumstances."

The crippled aircraft was shipped promptly to Lockheed in California for repairs. Now it was decided to reverse the global flight's direction, west to east. With leave from his liner about to expire, Captain Manning had to bow out as navigator. Fred Noonan would stay on. In record time, less than two months after the ground-looping mishap at Luke Field, the Electra was poised for flight again. On May 18th A.E. tested the plane and found it satisfactory. On May 19th she and her winged steed were reunited in Oakland, where Fred Noonan waited.

On May 20th the runway dropped out from under the Electra as Amelia headed it eastward. Ensuing days saw stops in Tucson (fire in the port engine after refueling, minor repairs), New Orleans and Miami. In Florida's queen city there was a pause of about a week for final adjustments and checks. On June 1st, NR 16020 was winging to Puerto Rico.

Subsequently logged were brief stops in Venezuela, Dutch Guiana, than Natal, Brazil, for fuel to span the South Atlantic. On June 7th they touched down in Senegal, French West Africa, followed by brief stops in French Equatorial Africa, Anglo-Egyptian Sudan, and Eritrea (Ethiopia). By mid-June the aerial expedition was in India after hopping across the Red Sea and Arabian Sea, a 1,950-mile

217

segment from Assab, Eritrea. Then came Burma, Singapore, Siam (Thailand today), Java, the island of Timor in Indonesia, Port Darwin in Australia, and Lae, New Guinea.

In Lae the stage was set for the grand finale, the baffling disappearance of Amelia Earhart and Fred Noonan.

Flight into Mystery

Approximately 7,000 miles remained to complete girdling the globe. It was appreciably less than the distance already covered, but way out yonder stretched that segment to tiny Howland Island. With its 2,600 miles of open Pacific it loomed as one of the most perilous legs of all. Certainly it would tolerate no errors in navigation.

Runways had been constructed by the U.S. Government on Howland, ostensibly to accommodate this history-making flight. Amelia and Fred must have drawn comfort from the knowledge that the U.S.S. *Ontario* was on station about halfway between Lae and Howland Island, the U.S.S. *Swan* had taken up a position roughly equidistant between Howland and Hawaii, and the U.S. Coast Guard ship *Itasca* had been ordered to stand by in the Howland area to serve as a radio homing-in beacon with her sophisticated equipment.

Threatening weather postponed takeoff from Lae on July 1st. Next morning, literally loaded to the gunnels with fuel, the Electra rose gracefully from the Lae airstrip and soon became a fading speck in the distance. Then that far sky was empty.

In a matter of hours all skies were to be empty of Amelia Earhart and Fred Noonan forevermore. Somewhere in the limitless monotones of sea and sky the Electra and its valiant pair vanished without a trace. That is, they disappeared without any *known positive* trace. What really happened to them is still being debated.

When word of the vanished plane reached the United States it knocked everything else out of newspaper headlines, stunning millions of disbelieving Americans. "Amelia Earhart lost? Impossible!" was the public's reaction. Many were confident that she had landed on a remote island somewhere and would turn up. Other news items faded into the background temporarily as people all over the world discussed the disappearance of Amelia Earhart and Fred Noonan.

One of the greatest, most intensive and costliest sea-air searches in history was launched. On direct request from President Franklin D. Roosevelt the U.S. Navy dispatched the aircraft carrier *Lexington*, battleship *Colorado*, and destroyers to the region where it was thought the Electra might have gone down. More than 250,000 square miles of Pacific Ocean were scouted on the surface and from aloft. Nothing.

So began a puzzle that three decades have failed to solve.

When, Where?

The simplest, most logical explanation is that NR 16020 carried A.E. and her navigator in a fatal plunge into the ocean. Any of several things could have been the cause: Freakishly severe weather; engine difficulties—although it seems improbable that both engines could have failed; or the miscalculation of the location of tiny Howland Island and the running out of fuel. Some observers considered that last a distinct possibility. Fred Noonan rated as one of aviation's best navigators; but his was a celestial system, dependent upon stars to determine positions, and it was believed that the Electra may have run into a spell of bad weather in which a high overcast prevented ''shooting the stars.'' An earlier message from a U.S. patrol plane, then about 420 miles north of Howland Island, reported a period of extremely bad weather between 2,000 and 12,000 feet in that region. If the Electra had overshot the island it might well have encountered the same sleet and snow reported by the seaplane, and been forced down or run out of fuel.

The Electra flew out of Lae on July 2nd and was due at Howland Island the next morning, where it would still be July 2nd because the isle lies east of the International Date Line. Amelia had given her estimated time for the flight as approximately eighteen hours. However, radio contacts with the Coast Guard's *Itasca* on station near Howland indicated that the plane was in the air at least twenty four hours and twenty-five minutes, underscoring a question as regards its remaining fuel supply.

Radio messages from KHAQQ (the Electra's call letters) were received at various late stages in flight by the *Itasca*. In an early-morning transmission, July 2nd (Howland time), fifteen minutes before Amelia's original estimated arrival time at the island, KHAQQ reported ''200 miles out and no landfall.'' That was followed in 31 minutes by ''Approximately 100 miles from *Itasca*, position doubtful.'' Later, but still early that morning, the plane reported 30 minutes of fuel left, no land in sight, and its position still doubtful. Sixteen minutes later, Amelia radioed that they were circling in an effort to pick up Howland Island. Along about then KHAQQ's signals were received in greatest strength by the *Itasca*. Tragically, due to apparent transmission problems in the Electra, neither the *Itasca*'s direction-finding gear nor the Navy's radio direction-finder on Howland was able to get a bearing on the plane.

219

Accounts of Amelia Earhart's last words in radio contact with the Coast Guard ship differ somewhat in their phrasing but agree generally in content: "We are in line of position 157-337 [degrees]. Will repeat this message. Will repeat this message on 6210 KC. Wait listening on 6210. We are running north and south." After that, silence.

Judging by signal-reception strength, the Electra was closest to Howland at about 0758 hours. KHAQQ mentioned a 157- to 337-degree "line of position," but navigator Noonan apparently wasn't sure where they were, and the *Itasca* was helpless to tell him. The final message from the aircraft reportedly was logged at 0844, Howland time, and calculations figured that the Electra ran out of fuel very soon afterward.

On July 19, 1937, the U.S. Navy officially suspended its search for Amelia Earhart and Fred Noonan.

Why, How?

As mentioned earlier, the simplest, most plausible theory to explain the Earhart-Noonan disappearance is that their Electra plunged into the Pacific, due to one cause or another. However, investigators* working on their own have spotlighted details and questions which vastly complicate the picture—and also make it much more fascinating. Because they are so intricate, these ramifications must of necessity be simplified here.

Among those conjectures offered, certain details assume prime importance: At the time of the ill-fated circumglobal flight the United States was technically at peace with Japan. However, Japan was already geared for war, and the master blueprint included the U.S. Certain Japanese-occupied key islands—Saipan, Tarawa, and Truk among them—were being readied quietly as stepping stones across the Pacific toward Hawaii. Should any of those preparations be detected by outsiders, Japan's plans would be out in the open, erasing any chance for surprise. Further, the country would lose face internationally, since such operations were in violation of League of Nations agreements. Accordingly, the islands involved were kept strictly off-limits to foreigners.

The U.S. suspected some kind of bellicose chicanery going on

---

* Prominent among these are writers Fred Goerner and Joe Klaas. Both have conducted long and incredibly involved searches for an answer to the question of what actually happened to the missing flyers. Highly recommended, if you want to go into the story in depth, are their books, Goerner's *The Search for Amelia Earhart* (Doubleday & Company, 1966) and Klaas's *Amelia Earhart Lives* (McGraw-Hill Book Company, 1970).

secretly on Japanese-occupied Pacific islands. But with permission to visit them being denied firmly, attempts to take a look might easily precipitate war. In 1937, the United States was in no position to lock horns with the Land of the Rising Sun.

Such was the situation that gave rise to the theory that a secret observation-photographic mission may have been the real reason for the Earhart-Noonan flight around the world.

This theory has engendered a number of questions and complications. For example, why would tiny Howland Island, with its hazardous runway infested by clouds of birds, be selected as a refueling stop when Canton Island in the Phoenix group offered a long Pan American Airways runway that would have been much easier to find? Howland Island was the flight's announced target, but theorists have pointed out that maybe Canton was the real destination, since a course to that island would have provided opportunities to flight-photograph a number of Japanese installations. That observation spawns another question: If Howland was the destination after Lae, why did the U.S.S. *Swan* and U.S.S. *Colorado* head for Canton Island in the Phoenix group to begin their search when the plane was reported overdue?

KHAQQ's last transmissions indicated that the Electra missed Howland Island and had lost its way. This wasn't consistent with Noonan's reputation as a crackerjack navigator. One angle of the photographic mission theory has suggested that he really knew where he was all the time, but, since Howland wasn't their target, he didn't want to reveal the plane's then-secret position. An inference is that vague or confusing reports were radioed to delude the Japanese and prevent possible interception by their aircraft. Fear of betrayal of the mission also was suggested as the reason for the Electra pinpointing its position only once during the flight's main segment, and then when only 800 miles out of Lae. The Japanese reportedly had had a small task force, which included carrier-based planes, waiting at Jaluit in the Marshall Islands ever since the around-the-world flight began. Were they routinely—or especially—suspicious?

Dim Alleys

One theory has proposed that, instead of the publicized Lockheed Electra 10-E, Amelia Earhart and Fred Noonan actually flew a supercharged, pressurized, higher-flying, long-range experimental Air Force plane, a Lockheed XC-35, more suitable to a secret photographic mission. This plane was being put through its paces over the Mojave Desert while Amelia and Fred awaited repairs on their Electra. Reading between the lines of this conjecture, one could infer that the ground "accident" at Luke Field in Hawaii was engineered to provide an opportunity for a switch. If so, however, the details of where and when the switch was made are publicly unknown.

You'll recall that the Electra's federal "license plate" was NR 16020. Another intent independent investigator, former Air Force major Joseph Gervais, was digging through Federal Aviation Authority records when he made a startling discovery. The registration number NR 16020 also had been issued to another Lockheed Electra, supposedly a modified version of that flown by Amelia Earhart and classified by its builder as model 12-A. The "R" in the 12-A's registration eventually was dropped, giving it the number N 16020; but the strange part of it was that the government supposedly never issued the same number to two aircraft. On the other hand, a plane could be repaired, like the Earhart-Noonan Electra, or be rebuilt— or modified, like the 12-A—and retain the same number.

The second Electra was of the same vintage as Amelia Earhart's. Investigator Gervais could find no records of transfer of ownership of the 12-A between 1937 and 1940, the period during which Amelia took possession of her model 10-E and subsequently winged into oblivion. But then, starting in 1940, the other Electra had a procession of owners. Interestingly, one of them was Paul Mantz, the globe-circling flight's technical advisor. He kept the 12-A for several years, and when he sold it in 1961 the plane was twenty-four years old. In December that year N 16020 crashed on a California mountain, killing two men aboard. With many years' experience as an Air Force crash investigator, Gervais examined the wreckage. A plate on the 12-A's manifold showed a delivery date of May 13, 1937, which was a week before Amelia Earhart and Fred Noonan took off from California to Miami on the start of their circumglobal adventure.

It's said that Paul Mantz once hinted cryptically that he knew the true fate of Amelia Earhart. Unfortunately, he never got the chance to prove it. Mantz did a lot of specialized and often dangerous flying for motion pictures. In 1965 he was killed in a crash during filming of *The Flight of the Phoenix*.

The Parade of Theories Marches On

Passing years add conjectures to those already offered in efforts to explain the disappearance of Amelia Earhart. The solution is still up for grabs. Take your pick of theories . . . you could be right.

1. They were on a bona fide around-the-world flight (with or without a photography assignment) and did indeed head for Howland Island. (If photography were part of the flight, a thought occurs

222

that perhaps the films were to be turned over to Navy hands at Howland.) Because of weather interference with navigation, they became lost, missed the island, and plunged into the sea when their fuel was exhausted. An independent investigator holding with this theory maintains that their plane's remains will be found in very deep water off the island.

2. The Electra's original engines were secretly replaced with more powerful Pratt & Whitneys for greater speed and higher altitude on an observation-photographic mission, after which the plane headed for Howland. They ran into bad weather that prevented Noonan from taking the required star sights and they were still too far from Howland to request bearings for fear of betraying the mission, or at least too far for the *Itasca* to be able to help them.

Or they miscalculated the head winds they were bucking, and actually were much farther from Howland Island than they thought. Soon a dwindling fuel supply placed them in a very precarious position. Believing they had overshot Howland, they overtook a prearranged emergency plan to head for some U.S.-held islands where they could make a beach landing. But because of the mistaken notion that they had overshot Howland, they flew away from the islands instead of toward them.

3. They were on a photographic mission which had one of these three endings: The Electra (or XC-35) was shot down by a Japanese interceptor; it crash-landed on or near a Japanese-occupied island, with either or both flyers being killed or taken prisoner by a Nipponese fishing fleet or warship; or the original plan was to land secretly at a prearranged remote island, whereupon they would be announced to the world as lost, then secretly rescued later, but something went awry and they never made it.

4. Often mentioned is the possibility that Amelia Earhart and Fred Noonan survived a forced landing on or near Japanese-occupied islands and were taken prisoner, perhaps to Saipan or Jaluit or even Japan. On Saipan, natives later told about seeing two American flyers, one with bandaged head injuries and the other with close-cropped hair like a man's (Amelia's hair was bobbed), taken prisoner. Their accounts variously held that the man was executed, either by decapitation or shooting, and the woman died in prison of dysentery; or both were executed; or either or both were whisked away, to where, the natives didn't know. One researcher made trips to Saipan to interview natives on islands that were Japanese-occupied at the time of the disappearance. He found many of them reluctant to talk, still fearful of Japanese reprisal after all those years.

Fragments of information gathered from natives seemed to point

223

to the two flyers being buried on Saipan. The researcher finally located an old, neglected cemetery where it was thought their remains lay. He managed to disinter some bones, which were brought to the U.S. for study by a competent anthropologist-anatomist. He pronounced them the bones of native islanders.

Along the same lines, it also has been suggested that the remains of Amelia Earhart and Fred Noonan were secretly recovered long ago, then placed in federal custody, very quietly, in keeping diplomatically with the disguise of the true reason for their around-the-world junket. Time and place of recovery couldn't be specified, of course.

5. In this theory they crash-landed on a lagoon on a remote atoll in the Marshall Islands, a territory then under mandate to Japan. Noonan sustained head injuries in the landing, but Amelia's luck in crashes held and she escaped uninjured. Natives brought Noonan ashore. During the days following, word leaked out somehow that the flight had been en route to Howland or some other destination from Truk, and the Japanese had a pretty good idea why. Then began a race between U.S. and Japanese naval vessels to get to them first. It had to be a one-sided contest, since the Marshalls were off-limits to foreigners. Possibly they were picked up first by a Japanese fishing boat. In any event, they were taken by a naval vessel to Japan's Pacific military headquarters on Saipan.

If the Earhart-Noonan flight was indeed a photographic mission, the flyers could have had the status of spies in Japanese eyes. And if their captors were aware of such a mission, they could have undergone an ordeal that culminated in death, either by execution or from hardship in prison. Under those circumstances, the U.S. could do nothing to save them. Any attempt would have been an open admission of guilt; and with Japan then possessing a military advantage it was no time to provoke a confrontation.

Most Fascinating of All . . .

There is the speculation that Amelia Earhart and her navigator did not perish, but did, in fact, secretly return to the United States. This conjecture naturally presupposes that the widely publicized report of their loss at sea was—as Mark Twain labeled the famous obituary of him which appeared while he still lived—''greatly exaggerated.'' Unlike the Twain obituary, though, the Earhart-Noonan report was no accident, but was purposely circulated as part of the mission's cover-up. Or so says the theory. Actually they were hiding out at a prearranged rendezvous, from whence they were secretly returned to the States. Back home, this speculation continues, they assumed different identities and quietly faded into the obscurity of private life. Beliefs have been expressed that Amelia Earhart and Fred Noonan were alive at least until 1960—and that one or both may still be alive somewhere in the U.S.

There has been nothing concrete yet to support the foregoing. However, one researcher tracked down a lady—she shall remain nameless here—who was an aviatrix of long experience, who knew many of the flyers A.E. did, and who even looked as Amelia Earhart might have that many years later. The lady denied publicly and vehemently that she was Amelia Earhart. (The question arises: Couldn't there have been a comparison of fingerprints?) Then there was the mysterious figure of a man who, because of certain physical features, could have passed for Fred Noonan at that age. Further, it seems that this fellow, like Fred Noonan, had worked for Pan American as a navigator at one time.

Debaters of the survival theory trot out other details. Amelia Earhart was declared legally dead in 1939, only a year and a half after her disappearance, whereas the usual waiting period is seven years. The inference, of course, is that the declaration was part of a cover-up. Also, there were some observers who believed—wishful thinking, perhaps—that the Japanese would not have dared to bring about the death of anyone as prominent internationally as Amelia Earhart. (Whether or not such immunity would have extended to Fred Noonan isn't stated.)

Bits and Pieces

A grim game, amusing in retrospect, was played in the Pacific during and after the search for the missing flyers. The Japanese were intrigued by U.S. activities on certain islands, including Howland. No sooner were Earhart and Noonan reported missing than the Japanese Government offered to aid in the search, and soon had an aircraft carrier and other naval ships under way toward Howland. It was believed that their "search" gave them an opportunity to inspect the Gilberts, Phoenix group, and other U.S.-held islands.

After the search was officially called off on July 19, 1937, the U.S. Navy Department asked permission of the Japanese Navy Ministry to have a ship, or ships, visit the islands of Palau, Truk, and Saipan, ostensibly to look for the missing plane but probably for some inspection. The requests were refused.

A 1960 biography of Amelia Earhart told how another famous aviatrix, Jacqueline Cochran, was sent to Tokyo right after Japan's surrender, becoming one of the first American civilians from outside the country to enter that city after World War II. According to the book, her stated mission was to learn about the roles played by

225

Japanese women in aviation during the war. If that were her purpose, why had she not done so during the war? An inference was that Jacqueline Cochran went to Tokyo to try to find out if Earhart and Noonan were still alive and prisoners in Japan as believed in some quarters. While there she reportedly came across files on Amelia in the Imperial Air Force headquarters, which files seem to have disappeared afterward. Reading between the lines, American authorities deemed it expedient to remove those records to preserve the secrecy of the Earhart-Noonan flight's mission.

Similarly, other Japanese military records which might have shed some light on their fate—and disposal, if dead—were removed from Saipan by the Japanese before the war's end. A suggestion was that those files contained incriminating evidence, and that, with defeat then in sight, the enemy had sudden misgivings about being responsible for anything happening to someone as universally well known and admired as Amelia Earhart.

Some bizarre threads run through the case. One is the story of a mysterious man named Wilbur Rothar (or Rokar), traced by investigator Joe Klaas (*Amelia Earhart Lives*). These are highlights:

Later in the summer of Amelia Earhart's disappearance, Rothar, a Bronx janitor, arranged a New York meeting with her husband George P. Putnam on the pretext that he possessed information about the missing aviatrix. Using the alias Johnson, he told a wild story. He was a member of the crew of a gun-running ship, he said, and when this vessel was only a few days out of New Guinea she found a wrecked aircraft on a lonely island. Lying on the plane's wing was the body of a man (Noonan), badly mangled by sharks. On shore was a woman. Rothar's ship buried the man's remains and took the woman aboard. Injured, emaciated, and out of her mind, the woman was treated by the vessel's surgeon and recognized from newspaper pictures as Amelia Earhart. This woman was gravely ill aboard his ship (not identified), said Johnson-Rothar, and he feared that desperate men in the crew would toss her overboard so that their illegal gun-running activities would not be exposed.

In August, 1937, Rothar was indicted for attempting to extort two thousand dollars from George P. Putnam. After sanity tests in Bellevue Hospital, New York, he subsequently spent more than twenty-five years in various mental hospitals, then finally dropped from sight completely.

During his internment Rothar was said to have babbled about a "boiler" having been blown up by ammunition. Any significance in this allegation stems from "Boiler" being an inside nickname of that Army Air Force XC-35, the experimental plane some observers thought Earhart and Noonan really flew. While a patient, Rothar announced repeatedly that he was Fred Noonan, former chief navigator for Pan American Airways (which Noonan had been) and navigator for Amelia Earhart on her around-the-world flight. Amelia was a prisoner of the Emperor of Japan, he added.

All these observations were in glaring contrast to what he told her husband. Yet there's another strange angle. Among psychiatrists' notes from hospital interviews was a crude map drawn by Rothar to represent the island where he said his ship had found the wrecked plane. On the map an examiner had scribbled the label "Harl Island," with figures that could be interpreted as a compass course and distance from Howland Island. It was suggested that "Harl Island" could have been the psychiatrist's misunderstanding of the name Hull Island, in the Phoenix group, a place where some researchers thought the Earhart-Noonan flight really ended. Oddly, Rothar's map and the doctor's notations fitted Hull Island.

Along about the autumn of 1962 Rothar disappeared, his whereabouts unknown.

Never an Answer?

Unless the simplest explanation—that the Electra 10-E plunged into the Pacific during a bona fide circumglobal flight—is accepted, the disappearance of Amelia Earhart and Fred Noonan remains a mystery. The passage of years has failed to write a conclusive closing chapter to their story. Perhaps irrefutable evidence will be uncovered somewhere, sometime; but I for one would hate to count on it.

I'd like to think that Amelia and Fred are indeed alive and well, cloaked in new identities. As of the United States Bicentennial in 1976 Amelia would be 78 years of age. That's not too old for her to watch with interest, and perhaps some amusement, the persistent efforts to find out what happened to her.

One thing is for sure: If she and Fred Noonan did go on a photographic mission during those critical prewar years, however it ended, their country owes them a great debt of gratitude. And it seems to me that if they survived, still more gratitude is owed for the sacrifices they made by fading into obscurity.

29

A Legendary Sea of Doom

Long ago, when kids could see a matinee for a dime and there was such a thing as penny candy, we tadpoles sat open-mouthed in the local Bijou watching a movie with a title something like *The Isle of Lost Ships*. Its locale was that far-flung expanse of central Atlantic Ocean called the Sargasso Sea, a huge region deriving its name from a kind of marine vegetation, sargassum weed, which collects there in great abundance.

Premise of that long-ago motion picture was that the Sargasso Sea is an expanse of weed-choked ocean in the vortex of a gigantic swirl of currents which ultimately transport everything they carry into the sea of weed, where an enormous collection of flotsam becomes trapped forever. You probably have guessed what the film was about; through centuries, disabled or becalmed vessels of many types, including primitive craft of ancient Phoenicians, squareriggers, and modern steamships, were carried into the weedy trap, where there was no escape. Within the vortex these entrapped vessels formed a large floating island. Its older vessels were tenanted by colonies of skeletons, the remains of earlier victims. Later arrivals made the best of their fate and lived out their lives in a sea-borne community, subsisting on fish, marine plants, and rain water. At least two of the inhabitants, celluloid attractions at the time, were sufficiently unconcerned about their predicament to have a romance.

Happily for us, we kids didn't worry about technical accuracy. Most of *The Isle of Lost Ships* was fiction, and other portions were based on ancient legends, Yet some truth also ran through the film.

And updating that old motion picture is the fact that the Sargasso Sea encompasses at least part of the seemingly deadly region we now call the Bermuda Triangle.

First, a Little Oceanography

My friend Dr. F. G. Walton Smith of the International Oceanographic Foundation and Dean Emeritus of the University of Miami's Rosenstiel School of Marine and Atmospheric Science, is one of the world's foremost oceanographers and marine biologists, as well as a leading authority on the Gulf Stream. In his definitive book about the Stream, *The Ocean River*, written with historian-anthropologist Henry Chapin, Dr. Smith points out that the world's oceans are gigantic machines, with wind, currents, sun, moon, and climates being factors in the machines' incredibly enormous exchange and output of energy. He likens each ocean to a wheel, continuously turning. North of the Equator, an oceanic wheel's movement is clockwise. South of the Equator, a wheel's movement is counterclockwise. This difference is due to the earth's rotation and is known scientifically as the Coriolis Effect, named for its discoverer, French mathematician Gaspard de Coriolis.

Oceanic wheels are, in effect, great currents, constantly in motion. The Gulf Stream is but one of many around our globe. Flowing northward from the Straits of Florida between the U.S. mainland and the Bahamas, it progresses offshore of the U.S. Atlantic seaboard. Approximately opposite New England and Canada's Maritime Provinces it obeys the laws of the earth's rotation and swings northeastward, then eastward, in clockwise rotation. Across the Atlantic Ocean it moves, now technically becoming the North Atlantic Drift, so called because a major moving force of the current here is the play of winds on the ocean's surface, creating a drift. Off the shores of western Europe* the flow again responds to the earth's rotation and, also influenced by the continent's mass in its way, continues its clockwise direction to veer southeastward, then southward. In this part of the Atlantic, offshore of the Iberian Peninsula, it's often called the Portuguese Current. Farther south an offshoot moves toward the northwestern part of Africa and is labeled Canaries Current. With northwestern Africa's mass bulging outward, the main flow once more obeys the Coriolis Effect, turning southwestward, then on a westward heading. Crossing the Atlantic Ocean, it becomes the North Equatorial Current. As it nears completion of its great circle, well off the southeastern United States it is frequently called the Antilles Current. The Antilles Current joins the Florida Current to form the Gulf Stream, and so the circuit is finished.

Roughly at the hub of this gigantic wheel of water lies the Sargasso Sea. Its outer limits and shape vary somewhat, due to fluctuations in the courses of the currents surrounding it, but it's crudely circular or elliptical in outline and sprawls across thousands of square miles in the Atlantic between the United States and

*This current's warming effect keeps some northern European harbors open in winter that otherwise would be closed by ice. It also inspires a species of palm tree to grow in southern Ireland. When I first heard about those trees I found it hard to believe until I saw them.

western Africa, extending almost as far eastward as the Azores and reaching southward nearly to the West Indies. Within this vast hub currents rotate more slowly; movement of water is sluggish. And there are doldrums, areas with little or no wind. Sargassum weed is transported here in great profusion by currents. When dense it gives the sea a brownish golden color, and it harbors a world of marine life all its own, many of whose adapted forms ride in with drifting weed. As torpid and windless as this enormous, weedy reservoir or eddy becomes, around it ocean currents move an estimated seventy-five million *tons* of water *per second* in clockwise rotation.

Now, Some Legends

The Phoenicians, great merchant-adventurers, and ancient Greek navigators were aware of those currents, referring to them as an ocean river, powerful and feared. From it there could be no return, they believed. Later the more fearless navigators ventured farther and farther out on the Atlantic. They may have authored the legends about the Sea of Coagulation or Coagulated Sea, pictured as a terrifying region which clutched vessels and held them in a grip of doom. Their Sea of Coagulation may very well have been the Sargasso Sea. Twenty-five hundred years ago a Carthaginian navigator wrote about an ocean region in which there was no wind and where the water was so choked with seaweed it held vessels back. He added that it teemed with many kinds of large, fierce, frightening sea monsters.

In 1436, a European mapmaker named Andrea Bianca prepared a chart of the Atlantic Ocean as visualized in that unenlightened age. On his chart he labeled the Sargasso Sea as *Mer de Baga*, "Sea of Berries," probably because the tiny air bladders which make sargassum weed buoyant could be mistaken for berries. It was an innocuous-sounding name, belying the dread in which the ocean west of Europe was held. Sea of Darkness it was called. Voyages into that region were strictly one-way trips, ancient mariners believed.

Christopher Columbus was a Genoese navigator who had learned his profession by sailing from Iceland to the Canary Islands off the northwestern coast of Africa (he probably was aided by that aforementioned Canaries Current). By then wiser scholars expounded the idea that the world was round, not flat like a table, beyond whose edge foolhardy mariners dropped into an eternity of space. Nevertheless, Columbus and other navigators still feared the Sea of Darkness, whose limits and nature were largely unknown. Many kinds of horrors were attributed to the Sea of Darkness, and possibly it held even greater terrors. No one really knew if all this was fact or fancy, and even the boldest mariners were understandably reluctant to find out. In light of such belief, Columbus's courage becomes phenomenal.

Wooden derelicts have been current-carried into the Sargasso Sea's sluggish vortex, there to rot and be slowly riddled by teredos or shipworms until they disintegrated. It's still possible, theoretically at least, that sails-only craft could go astray and be transported into the Sargasso's "horse-latitudes" (windless areas), and there get becalmed in isolation until water and provisions ran out. For all we know, it may have happened many times over the years.

Unfortunately for the credibility of that long-ago movie about the island of lost ships, however, Columbus proved that the vast reservoir of the Sargasso Sea is not so choked with weeds as to imprison vessels venturing into it. Even so, for centuries after his historic voyage to the New World it remained a popular belief among mariners that the Sargasso Sea is a floating cemetery of vessels that had collected there since remote times. Maybe Hollywood producers believed it too.

And Yet

There *is* an eerie parallel between some of the ancients' beliefs and what has since transpired in the Sargasso Sea and environs, notably that place now called the Bermuda Triangle. Although it's not quite the same as that described by ancient mariners as their Sea of Coagulation, there does seem to be a kind of "clutching" in the Triangle. In that sense, many vessels, along with aircraft, have met grief within its precincts. But beyond that, ancient and modern concepts part company. Although we still do not know for certain why some vessels and airplanes have vanished in the Bermuda Triangle, we do know that they were not trapped by a sea of weed.

30

Revisiting the Bermuda Triangle

In recent years much attention has been focused on that region of western Atlantic Ocean now popularly known as the Bermuda Triangle. Under that name, or such eye-riveting aliases as "Devil's Triangle," "Deadly Triangle" or "Triangle of Doom," it has starred in several magazine pieces. At least half a dozen books have been devoted to it. Television documentaries have zeroed in on the Bermuda Triangle too. There's even a kids' game by that name.

It seems inevitable that Hollywood, with its penchant for capitalizing on disasters, will grind out motion pictures about the Bermuda Triangle—conceivably with a sequence in which a currently popular rock-music group caterwauls a number as their vessel or aircraft disappears beneath the surface. And how come someone hasn't created a stage play or musical based on the Triangle?

The name Bermuda Triangle caught on, but it's a misnomer on two counts. In the original concept the region was roughly triangular in outline. More recently, though, its reaches have been expanded until now it is variously described as a rectangle or, by U.S. Coast Guard definition, a gigantic semicircle. The "Bermuda" in its name gives an impression that the region is more or less restricted to waters around that charming island, whereas it actually sprawls across an immense region of many thousands of square miles. But for our purposes the misnomer is not a drawback, so I'll go with the popular "Bermuda Triangle."

Background

Dating back to the earliest North American colonization, and earlier, craft of many kinds have sailed in and across the Bermuda Triangle. Christopher Columbus sailed it 500 years ago. Florentine navigator Giovanni da Verrazano negotiated part of it when he explored the continent's Atlantic seaboard in 1524.

Uncountable pleasure cruises into and through the "danger zone" have been logged by liners steaming from New York to Bermuda, the Bahamas, and the Caribbean isles. For years, commuting by the Furness Line's *Monarch of Bermuda* was a prominent part of the U.S. travel scene. During the 1920s and 1930s the late Ward Line operated a shuttle service between New York and Havana, with one of its cruise ships in the Bermuda Triangle every few days. Literally thousands of sail-powered merchantmen, freighters, and tankers have plodded through Triangle waters to and from Boston, New York, Philadelphia, Baltimore, and ports to the south. Countless sport and commercial fishing boats have navigated in the region. The late Ernest Hemingway, an ardent big-game angler, took his *Pilar* into Triangle waters in search of blue marlin.

And who could possibly count the number of commercial airline and private-plane flights that have soared over the Triangle or crossed it to its fullest? I've flown over it many times, in trips to Bermuda, the Bahamas, Florida, Virgin Islands, Jamaica, Puerto Rico, and other destinations in the Caribbean Sea.

Point is, all those excursions into, across, and over the mysterious region have been made safely, logging millions of passenger-miles. I know I wasn't lost in the Bermuda Triangle.

However it's also a fact that numerous vessels of various kinds, along with assorted aircraft, seem to have vanished without a trace or logical explanation in the Bermuda Triangle region during the past 100 to 150 years. In addition to ships of appreciable size, unexplained disappearances embrace many smaller craft, such as fishing vessels and private boats. Among the latter, possibly, was the famous sailboat *Spray*, manned single-handedly by her owner, Captain Joshua Slocum. In recent decades the Bermuda Triangle's log of the missing has lengthened considerably with the addition of aircraft—government, commercial, and private. And, in not the least of all the infamous region's puzzles, people have disappeared from vessels without clues to their fate.

The subject of the region's mysteries and danger has become controversial. On one side are substantiated occurrences which can't be ignored, along with credible accounts and considered opinions of knowledgeable observers. In the other faction stand the skeptics. At least one book and a television program have "debunked" the Bermuda Triangle as a region of menace. There are knowledgeable observers on this side too. Yet even some of the skeptics will admit that there are some pretty strange goings-on in that oceanic region.

Believers have offered a multitude of theories to explain the vanishing of vessels and planes without a trace. These include: an as yet unknown magnetic or electrical phenomenon; complex time-space warps (too intricate to detail here); weird holes in the ocean floor with an extremely powerful "suction" force; visitors or raiders from outer space—UFOs; people or human-like creatures that live on or under the ocean floor and come to the surface in vehicles of their own; and natural or supernatural forces as yet unknown. Skeptics and disbelievers are inclined to scoff at most of these speculations, holding firmly with a belief that there are logical, plausible explanations as yet uncovered.

The Testimony

Whatever the causes—logical, unexplainable, or supernatural—the Bermuda Triangle expanse probably has been claiming vessels ever since man first ventured on its surface. Only within the past couple of decades have we begun to take a harder, closer look at the region and list its casualties. This has prompted investigators to do some retrospection, and in that they discovered unsolved mysteries dating back more than 150 years. Now those disappearances are generally included in the Bermuda Triangle affair, although it's to be wondered if sometimes the region's limits aren't stretched to accommodate them. We won't argue that point here. Let's assume that they did occur within the section of Atlantic Ocean now defined as the Bermuda Triangle. The trouble is, information concerning some of those old mysteries is only fragmentary. We can find no notations about weather and sea conditions at the time, factors which might very well lift disappearances out of the mystery category.

If accounts are correct, 1800 was a bad year for the U.S. Navy in the Triangle region. Two ships were lost. The U.S.S. *Pickering* vanished with ninety officers and sailors while on a presumably routine run from Delaware to the West Indies. The other disappearing act claimed many more victims. In August, 1800, the U.S.S. *Insurgent* seemed to vanish into thin air with 340 men.

Mysterious disappearances during the next forty to fifty years included these: (1) French vessel *Rosalie*, 1840, found derelict and abandoned, with no traces of her crew (reported in the London *Times* that year). (2) Schooner *Bella* and men vanished, 1854. (3) March, 1854, *City of Glasgow*, on her way from Liverpool to Philadelphia, carried approximately 450 people to who-knows-where (reported in a New York newspaper). (4) In 1870 the *City of Boston* entered the unknown during an easterly crossing from New York to Liverpool. With her evaporated 177 persons. (5) In January of 1880 came the perplexing, never-solved case of the British Navy's training ship H.M.S. *Atalanta*.

Justifiably spotlighted as one of the Bermuda Triangle's most baffling mysteries is the total disappearance of the U.S. Navy's supply ship *Cyclops*. She earned stellar billing in 1918.

Technically, the 500-foot, 19,000-ton *Cyclops* was a collier, a ship designed to transport coal. But she also carried ore, and on her final assignment her belly was pregnant with some 10,000 tons of manganese ore, used in manufacturing steel. Some of that steel, in turn, would find its way into weapons with which to fight Germany in the European war still raging. Also aboard the *Cyclops* were approximately 309 people—officers, crew, and U.S. Navy personnel.

On March 4, 1918, the *Cyclops* left Barbados in the British West Indies for Norfolk, Virginia. She steamed in fair weather, and no unusual sea conditions were reported. Somewhere along the way she, her people, and the manganese ore were completely swallowed. Deepening the mystery, no wireless messages were received from her; and, typical of Bermuda Triangle enigmas, she left no traces.

Considering that a state of war existed between the United States and Kaiser Wilhelm's forces, and that she was a U.S. Navy vessel, a logical speculation is that she was sunk by an enemy U-boat or surface raider. Or maybe she struck an errant mine. Two flaws in these theories are the complete absence of debris, bodies, and survivors, plus the fact that no radio message came from the ship. In either event it hardly seems possible that she could sink without her radioman being able to flash an SOS. Further, a search of German Imperial Navy records after World War I turned up no submarine, or evidence of mines, in that area at the time.

So far as was ever learned, the *Cyclops* didn't encounter a hurricane (March isn't the hurricane season thereabouts) or even a localized freak storm of exceptional savagery. The latter is possible, but one would think that her radio operator would have announced it. A U.S. Navy expert theorized that the *Cyclops* suddenly was flipped upside down and sank almost immediately. This would require an uncommonly great force, such as winds and seas of a severe hurricane, or—a possibility, although not reported—a tidal wave generated by an undersea earthquake (such waves reach unbelievable heights and velocities). Such a catastrophe would have to strike with unusual suddenness to prevent even one SOS from being transmitted. Too, it would seem that 10,000 tons of ore would help as a stabilizer. However, if the ship broke open or otherwise took in a lot of water, that ore would expedite her descent to the bottom.

Sabotage was a logical proposal, for the same reasons that the *Cyclops* could have been the target of an enemy war vessel. Feeding this theory was the fact that her captain, although a U.S. Navy man, was born in Germany and had changed his name from the Teutonic form Wichmann to an anglicized version, Worley. (There was nothing particularly sinister in the name-changing, it should be added. A number of loyal German-American citizens did it simply because of high public feeling against Germany.) Some probers also challenged Captain Worley's sanity. They figured that a skipper who allegedly walked his bridge clad in long underwear and a derby hat just might not be totally sane. In any case, nothing was established to link him with sabotage.

It was said that after the war German agents in South America claimed credit for eliminating *Cyclops*, professing to have secreted delayed-action bombs in her holds. That was plausible, but it turned out to be so much talk, never substantiated. And here too it seems unlikely that the ship would sink before her wireless operator could get off an SOS.

The U.S. Navy eventually rang down an official curtain on the case, admitting that the disappearance was one of the most baffling in its history. The Navy concluded by saying that not one of the many theories advanced explained it satisfactorily.

That is the status of the *Cyclops* mystery to this day.

Some Members of the Show's Supporting Cast

There are dozens. We'll look at some of relatively recent vintage.

Case 1: Gradually being elevated from "supporting cast" to "star" or classic status in the Bermuda Triangle drama is the 524-foot *Marine Sulphur Queen*, a World War II Type T-2 tanker converted to carrying molten sulphur. This was her cargo—more than 15,000 tons of it—when she and her crew of thirty-nine pulled out of Beaumont, Texas, bound for Norfolk, Virginia, on February 2, 1963. It should be mentioned that her owners did not consider molten sulphur an especially dangerous cargo.

On February 3rd a routine radio message from the *Marine Sulphur Queen* placed her well out in the Gulf of Mexico. The communiqué mentioned no problems; but during the day another tanker, about forty miles away, reported very heavy weather, with mounting seas and wind in increasing velocity. On February 4th came another message from the *Marine Sulphur Queen*, still with no word of trouble. By then the tanker should have been approaching the Dry Tortugas, islands belonging to Florida and offshore of that state's southernmost tip. After the February 4th message there was nothing but silence. Efforts to contact the *Marine Sulphur Queen* by radio failed.

On February 6th she was declared overdue, precipitating a massive sea-air search by the U.S. Coast Guard, Navy, Marines, and Air Force. The search fanned out from the missing tanker's last known position all the way to Virginia. To no avail. A scouting plane reported seeing a yellowish discoloration of the ocean, such as might have come from sulphur, off Jacksonville, Florida. Checking by a surface vessel revealed that the aircraft had spotted great masses of seaweed, probably sargassum weed, which has a golden, yellowish tinge. More than 300,000 square miles of water were searched. Further, the Coast Guard checked numerous ships that might have seen the *Marine Sulphur Queen* on February 4th or 5th. Here they also drew blanks. When a week of searching proved fruitless, the hunt was discontinued.

Days later a Navy vessel came across a patch of flotsam that included a life jacket bearing the tanker's name, about fifteen miles southeast of Key West, Florida. Further probing by a Coast Guard patrol boat turned up another labeled jacket. These findings launched a second search. This one retrieved an assortment of gear identified as having come from the *Marine Sulphur Queen*. Included were seven or eight life jackets, some preservers of the ring buoy type, one or two name boards, and other miscellaneous items. They provided no positive clue as to what had actually happened to the missing ship.

There's a very unusual angle to the case, in addition to the disappearance. According to accounts, it was a crew member's speculation in the stock market that provided a warning that all was not well with the *Marine Sulphur Queen*. Shortly before she left Beaumont, the crewman asked a brokerage firm to buy certain stocks for him. The brokers executed the purchase, then wired confirmation to the tanker, now at sea. When efforts to contact the vessel failed, her owners were notified, and the mystery of the *Marine Sulphur Queen* began.

As in the case of the U.S.S. *Cyclops*, no message of distress came from the tanker. Whatever fate overtook her, it kept her silent. Explosion of her cargo of molten sulphur—kept liquid in insulated tanks, at a temperature of about 265° F., by coils—was considered a possibility. Although her owners did not consider molten sulphur appreciably more dangerous than other cargos, it was suggested that agitation of the liquid during the storm many have released an unusually large volume of potentially explosive gases. Another possibility was that she broke apart during that February 4th storm.

Or perhaps a hull failure, caused by changes during modification, admitted water that created a steam explosion when it came in contact with molten sulphur. Sudden capsizing by freak-waves also was considered. Still other conjectures included hijacking, sabotage, hitting a stray mine, and capture by Cuban forces or pro-Cuba sympathizers. Those last proposals were discarded.

It was agreed that the *Marine Sulphur Queen* must have gone to the bottom in a hurry, whatever the cause. She transmitted no distress calls. No bodies, lifeboats or survivors were found. In connection with fast sinking, attention was drawn to the *Queen*'s construction. Type T-2 tankers and other ships have been known to break into halves under severe stress, but with watertight compartments their halves can remain afloat. It seems that several of the *Queen*'s watertight compartments, as well as certain hull-strengthening transverse bulkheads, had to be sacrificed during modification to make room for a 306-foot-long cargo tank. Experts believed that with such an arrangement she could sink quickly if an explosion or structural failure opened her hull or a storm broke her in two.

At the conclusion of its official probe the U.S. Coast Guard announced that a definite cause of the *Marine Sulphur Queen*'s loss could not be pinpointed. So the mystery is not *that* she vanished, but *how*.

Cases 2, 3, 4: These freighters became Bermuda Triangle mysteries during the period from 1925 to 1931: The *Cotopaxi*, bound for Havana out of Charleston, South Carolina; the *Suduffco*, somewhere south of Port Newark, New Jersey: the *Stavenger*, with 43 people aboard, vanished south of Cat Cay, Bahamas; and the *Anglo-Australian*, with her crew of 43, disappeared after an all's-well message while plodding along west of the Azores.

Case 5: Another freighter, the *Sandra*, pulled out of Savannah, Georgia, one June day in 1950 with a cargo destined for Puerto Cabello, Venezuela. After she passed St. Augustine, Florida—in fair weather—she was never heard from or seen again.

Case 6: There is the riddle of the *Anita*, a freighter in the 500-foot class, 13,000 tons, manned by a crew of 32. She left Norfolk, Virginia, on March 20, 1973, with a cargo of coal for West Germany. During the next couple of days she disappeared completely, without so much as a radio peep. The only trace was a ring-buoy-type life preserver with her name stenciled on it; and, as the Coast Guard pointed out, that could have dropped or been blown overboard earlier.

The *Anita* was half of a more or less simultaneous double loss.

238

The other half was the freighter *Norse Variant*, of the same type and size as the *Anita*. Both vessels were of Norwegian registry but had different owners. Coincidentally, the *Norse Variant* also was headed for West Germany with a cargo of coal, and also sailed from Norfolk. In fact, she departed scarcely two hours ahead of *Anita*.

On March 21st the Atlantic was convulsed by a very bad storm: Waves from thirty-five to forty-five feet high with wind gusts in excess of seventy knots. When about 150 miles southeast of Cape May, New Jersey, the *Norse Variant* reported that she was sinking and that her crew was taking to lifeboats (in such seas that wasn't much of an alternative). There was only one survivor. He told his rescuers that a sudden, extremely violent wind, accompanied by mountainous waves, struck the freighter. Huge seas swept across her deck, tearing a hatch cover free and pouring tons of water into her holds. She sank within a few minutes.

But there were distress messages from the *Norse Variant* and some identifiable debris was found. Nothing identifiable, except that one ring buoy, remained from the *Anita*. No bodies, no survivors. A plausible possibility is that she was another victim of the same storm, going down so fast that her radio operator couldn't transmit a "Mayday!" distress call. However, other ships made it through that storm safely, and maritime experts stated that the odds against two ships of the same type and size being sunk by the same storm are very high.

So what really happened to the *Anita*?

Case 7: If the Bermuda Triangle's limits should be extended to include the Gulf of Mexico's eastern reaches, as some probers suggest, a 1966 riddle has to be added. This one is really odd. On October 26th, the 67-foot tug *Southern Cities* chugged out of Freeport, Texas, towing a 210-foot barge laden with chemicals. When the tug's radio contacts ceased, search planes took to the air. They soon found the barge, its cargo undisturbed, but the tug and her crew had vanished without even a hint at their fate.

Smaller craft: Whatever answers there may be to the Bermuda Triangle's puzzling questions, there is an alarmingly lengthening list of missing smaller craft, commercial and pleasure.

The sport fishing boat *Sno' Boy*, with forty people aboard, vanished offshore of Kingston, Jamaica, in 1963. Debris, thought possibly to be from the missing boat, was seen; but her disappearance is still a mystery.

Another vanishing act was that of the 55-foot *Evangeline* and her people while bound for the Bahamas from Miami, 1962.

Dancing Feather, a 36-foot ketch, was added to the list of

Bermuda Triangle mysteries in 1964 when she seemingly evaporated somewhere between Nassau, Bahamas, and a destination in North Carolina.

Also missing and unaccounted for, the 56-foot schooner *Windfall*, off Bermuda, a 1962 entry.

The 58-foot, 20-ton schooner *Enchantress* vanished somewhere between Charleston, South Carolina, and St. Thomas, U.S. Virgin Islands, in January of 1964 during the early stages of a voyage to the home port of her new owner in California. With her went her owner, his wife, two young sons, and a veteran of single-handed transoceanic skippering, Count Christopher de Grabowski.

Of steel and wood construction, 1925 vintage, the good-looking *Enchantress* was along in years, but had been reconditioned at a Long Island yard prior to being taken over by her new owner. She carried a small dinghy and life jackets of an approved Coast Guard type. Her equipment also included a radio direction-finder and a powerful radiotelephone.

The *Enchantress* may not qualify as a full-fledged Bermuda Triangle mystery, however. She had encountered dirty weather off northeastern Florida, and in one of her "Mayday!" calls reported water knee-deep in her cabin. At the request of the Coast Guard's Jacksonville station the distressed boat began a long count on radio to provide an opportunity to fix her position, later calculated to be roughly 150 miles or so SSE of Charleston. In a poignant note, the count was being made by a boy's voice when it faded away in the last ever heard from the *Enchantress*.

An extensive search by Coast Guard and Navy surface vessels and aircraft turned up bits of debris that included a flotation cushion. It was thought that the cushion may have come from the missing schooner.

The disappearance of the *Enchantress* and her five passengers was probably more tragic than mysterious. From what is known of the case, heavy weather undoubtedly was a major contributing factor, with any of a number of disastrous things following. Continuous sweeping by large waves may have dislodged her hatch covers and filled her with water. The schooner had a wooden hull. A serious leak could have stemmed from failure of an old through-hull fitting, or from seams sprung by continuous pounding, or she may have struck something that punched a hole in her hull. Conceivably the leak went undetected until it was too late, and her pumps could no longer handle it.

All that seems certain is that the *Enchantress* sank on January 13, 1964. We'll never know why.

The big, posh, brand-new, and superbly equipped pleasure

cruiser *Saba Bank* was valued at more than a quarter of a million dollars. She had been built for a Delaware firm and was specially equipped for charter to scuba-diving groups. She vanished completely with her four-man crew sometime after last being heard from on March 24, 1974, while on a roundabout trip from Nassau, Bahamas, to her home port of Miami. Ironically, it was a shakedown cruise. *Saba Bank* was due in Miami on April 8th at the latest. When she failed to arrive by April 10th the Coast Guard listed her as unreported; and when a preliminary search by radio in Florida and the Bahamas failed to elicit responses, the classification was changed to overdue.

With both the U.S. Coast Guard and Bahamas Coast Guard participating, a search was begun on April 10th. It lasted for two weeks and reached as far south as Central America on the chance that the vessel may have been hijacked or stolen. No trace of the *Saba Bank* and the four men known to have been aboard was found. So far as is known, weather and sea conditions were not a factor; and her being new and radio-equipped only deepened the mystery. At last report her owners' widely posted offer of a $2,500 reward was still uncollected. The handsome *Saba Bank* seems to have become yet another Bermuda Triangle victim.

Valuable tips to ocean-going yachtsmen were gleaned from the U.S. Coast Guard's comments about this case. The search for the big cruiser was hampered seriously by lack of information about her proposed cruise route. It's always wise to file a navigation plan with someone on shore (even for just a day's excursion offshore, I might add). In light of a possible pirating or "yachtjacking," the Coast Guard warned that yachtsmen should always know everyone aboard and be very wary of strangers who want to hitch a ride or sign on as crew members. It has been reported that a number of private and commercial fishing craft have been hijacked or pirated, either in theft or for use in running narcotics.

In April 1975, a 73-foot shrimp boat named *Dawn* added her name to the Triangle's roster of mysteriously vanished vessels. Her home port was Key West, Florida, and she was earning her living in that state's east coast waters when she disappeared without a trace. With her went her owner-skipper and an equally experienced crew of two. The *Dawn* was last seen by another craft on April 22nd when only ten miles or so offshore, apparently in clement weather. After that, nothing: No radio messages, no flotsam, no survivors or bodies. A search by Navy, Coast Guard, and Air Force planes failed to find a single clue.

Accounts of sea conditions that night are contradictory (one says "calm seas"; others mention winds to twenty knots and four- to six-foot waves), but the story of the 23-foot cabin cruiser *Witch-craft* remains intriguing:

On the evening of December 22, 1967, the *Witchcraft*'s owner, a retired hotel man named Daniel Burack, invited his friend Father Patrick Horgan of Fort Lauderdale to view the Christmas lights of Miami Beach from seaward. At the farthest they would run only a mile or so out into the ocean for a panoramic view of the city.

At 2100 (9:00 P.M.) the Coast Guard in Miami received a radio communiqué from Burack announcing that his boat was mechanically disabled. Her propeller had been bent badly by striking a submerged object, and the resultant excessive vibration necessitated turning off her engine. Burack, a capable skipper, was calling to request a tow back to port. He apparently was not overly concerned about any waves. The cruiser's hull was undamaged, and with its built-in foam flotation was considered virtually unsinkable. Further, there were approved-type life preservers on board. Burack gave their approximate position as being in the vicinity of a Buoy No. 7. The Coast Guard advised him that help was on the way and asked him to fire a flare in about twenty minutes to guide the rescue vessel.

A Coast Guard boat was in the *Witchcraft*'s reported location in just under twenty minutes. No signal flare was seen, there had been no further radio communications, and the cruiser had vanished. A five-day sea-air search fanned out southward to Islamorada in the Florida Kays, north to St. Augustine, and 120 miles out on the Atlantic, covering approximately 25,000 square miles. In a way the *Witchcraft* lived up to her name. She and her two passengers had been made to disappear. No trace was ever found.

What makes this case unique in Bermuda Triangle annals is a disappearance so close to shore, practically within a few flaps of a seagull's wings from Miami Beach. Not easy to explain is why there were no further radio communiqués from the cruiser. Although she was helpless without her engine, it probably could have been used in an emergency to keep her headed into the waves if there appeared to be danger of broaching (being hit broadside and swamped). However, broaching looms as a distinct possibility, and if she were swamped but still afloat her radio could have been rendered useless. Also a distinct possibility, *Witchcraft* was not as "virtually unsinkable" as believed.

That she was not found at night in darkness and choppy seas isn't surprising, what with bad visibility, her small size, and the wind masking any shouts from her passengers. Nor would it have been easy to spot her even by day from the air. A 23-foot boat becomes a small object on a big ocean's expanse. Too, with shifts in the wind she conceivably could have been blown into the clutches of the northbound Gulf Stream and transported many miles from her original position—if she were still afloat.

In any event, the *Witchcraft* is catalogued among the unsolved mysteries of the Bermuda Triangle.

Another Kind of Mystery

In a number of Bermuda Triangle episodes the vessels remained and were discovered, but their people had vanished, without any clue to the reasons. Often these disappearances are more baffling than the loss of entire vessels. Here is a sampling of cases.

1. April 1932: Two-master sailing ship *John and Mary* found derelict and adrift, but *freshly painted*, no one on board, about fifty miles south of Bermuda. Reason never learned.

2. October 22, 1944: The *Rubicon*, a Cuban vessel, encountered adrift and abandoned off Key Largo, Florida, devoid of life except for a hungry dog. Her last log entry, dated September 26th, indicated she was then still in Havana. The lifeboats were missing, suggesting that an emergency at sea caused the abandonment. What is odd is that the dog was left behind. Whenever possible seamen take the vessel's mascot and their own pets with them when they leave a ship. Considering that World War II was on, one could wonder if her crew were removed forcibly by a prowling German submarine, although it's contrary to procedure that any enemy raider didn't sink the evidence. In a hair-raising theory it was suggested that maybe the *Rubicon*'s men were captured by visitors from outer space (collecting samples of life on this planet, perhaps?).

3. June 1969: There was nothing to indicate what had happened to those aboard the 60-foot *Maple Bank* when she was discovered, a wandering derelict, north of Bermuda.

4. July 1969: The 12-meter sailing yacht *Vagabond*, abandoned but sound, found drifting west of the Azores.

5. April 1925: Bound for Hamburg, Germany, from Boston, the Japanese freighter *Raifuku Maru* plodded into very bad weather and soon was in big trouble. SOS calls crackled from her wireless and were picked up by the White Star liner *Homeric*, which rushed with all possible speed to her assistance from about seventy miles away. The last message from the *Raifuku*'s radio operator was in fractured English, but it made a point: "Now very danger. Come quick." The *Homeric* reached the stricken freighter. In his report the liner's skipper said he found the *Raifuku Maru* a badly listing derelict (meaning no one aboard), with her lifeboats smashed beyond use. Despite stormy seas, the *Homeric* had hoped to pick up

survivors, but there were none; nor were any liferafts seen. If the freighter was derelict—abandoned, yet her lifeboats were smashed and there were no men in the water or on liferafts, the fate of her crew becomes a mystery, unsolved.

6. July 1969: In one of the weirdest Triangle items of all, a dispatch from the Reuter's news agency and an item in the London *Times* announced discovery of five vessels abandoned in the same general area in which the derelict *Mary Celeste* was found more than a century earlier. Three were sound and apparently riding in normal fashion; two had flipped over. There were no signs of any of their crews.

In the Air

As time goes on, the list of various kinds of aircraft—private, commercial and military—disappearing within the Bermuda Triangle lengthens and threatens to rival that of missing surface craft. What is particularly disconcerting here is that the presence of planes within the region is relatively brief when compared with that of surface vessels.

Now the most famous case, that of the U.S. Navy's five TBM Avenger torpedo bombers, datelined Fort Lauderdale, Florida, on December 5, 1945. Poised for a routine training flight from the Naval Air Station there, the five aircraft had been checked carefully. Everything—engines, controls, all instruments—was in perfect working order, according to testimony at the inquiry. All five Avengers had been fueled to capacity for a cruising range in excess of 1,000 miles. Survival gear aboard each included life jackets and an inflatable raft. On the Station's log this training mission would be Flight 19. Each plane normally was manned by a crew of three: Pilot, gunner, and radio operator.

Flight 19 took off at 1400 (2:00 P.M.) in brilliant Florida sunshine. Its mission lay along a triangular course: 160 miles east, then 40 miles north, then a southwest swing back to the base, all well within the Avengers' 1,000-plus mile range. So far as the Naval Air Station was concerned, the next important communiqué from Flight 19 would be an announcement of its estimated time of return.

Instead, there came a jolting message a little over an hour after takeoff. It was voiced by Lt. Charles Taylor, Flight 19's leader. The reported phraseology varies a bit, as does the time of its reception, but here's its substance: "This is an emergency . . . We can't see land . . . Repeat, we can't see land . . ."

When the base asked Flight 19's location, Lt. Taylor replied, "We're not sure of our position . . . We can't be sure just where we are . . ." Considering the pilots' experience, this was a surprising admission. More startling was the flight leader's answer when the Naval Air Station advised altering course to due west to bring them to the coast: "We don't know which way is west . . . Everything is wrong, strange . . . We can't be sure of any direction . . . Even the ocean doesn't look as it should . . ." In addition to that reference to the ocean, another mystery within the mystery is why, flying in clear weather, the pilots couldn't see the sun to fly west.

Soon after that, a message from an Avenger reported both compasses out. The pilot thought he might be over the Florida Keys—little islands strung like beads on highway U.S. 1 for 150 miles between Miami and Key West—southwest of Fort Lauderdale. He was advised to fly north to follow the Keys to the mainland. Not long afterward the same pilot radioed that he was passing over a small island, with no other land in sight. He could not have been over the numerous Keys. He was lost.

Communications between Flight 19 and Fort Lauderdale worsened. The aircraft evidently could no longer receive the base's transmissions. However, even though static hampered reception of messages from the mission, the Naval Air Station could still hear fragments of exchanges among the pilots. They were very strange and contained references to all the planes' gyro- and magnetic compasses "going crazy" and showing different headings. There were references to possibly running out of fuel, along with mention of winds to seventy-five miles an hour, which would be hurricane force. They took off in fair weather, remember.

As time wore on, the mystery deepened. At about 1600 Fort Lauderdale learned that for some reason flight leader Taylor had turned command over to another pilot, a U.S. Marines captain named Stiver. Subsequent communications from Stiver indicated growing confusion in Flight 19: Admission that the pilots didn't know where they were, a belief that they were 225 miles northeast of base, then a contradictory thought that they must have crossed Florida and were somewhere out over the Gulf of Mexico (difficult to fathom, in clear weather, with seasoned pilots). Soon afterward, messages indicated that the flight leader had elected to fly a 180-degree course, due west, in an effort to find the coast. However, his messages grew increasingly fainter, suggesting that his direction was 180-degrees out, and that they were flying east, away from the coast.

Reports differ when quoting Flight 19's final messages. One has it that at about 1625 (4:25 P.M.) Captain Stiver reported to the base,

"We don't know where we are . . . It looks like we are . . ." then silence. Other accounts add: "Entering white water . . . We are completely lost . . ."

With Flight 19 in trouble, rescue missions were launched quickly. One involved a Martin Mariner "flying boat" specially equipped for sea rescue work and carrying a crew of thirteen. It sped toward the last known position of Flight 19. For ten to fifteen minutes the Mariner maintained radio contact with its base. Then, with a report of strong winds above 6,000 feet, the flying boat's communications ceased abruptly. Shoreside efforts to contact it brought no response. Now the other search units were notified that the Mariner also was missing.

Six planes and nearly thirty men appeared to be Bermuda Triangle victims.

One of the most massive sea-air searches in history was undertaken. Participants included upwards of 300 government aircraft in a low-flying crisscrossing pattern, eighteen Coast Guard vessels, U.S. destroyers, and submarines, flying boats from the Banana River Naval Air Station, and volunteer private aircraft and vessels. From the Bahamas, British Navy and RAF units joined the search. Land parties combed Florida and Bahamas beaches for wreckage, survivors, and bodies. Nearly 400,000 square miles of sea, including the Caribbean and large sections of the Gulf of Mexico, were covered in detail. No survivors, bodies, wreckage flotsam, oil slicks, life jackets, or other clues were ever found. It was as though the five Avengers and the Mariner had changed into thin air.

Understandably, the Navy admitted it was baffled. Fact is, no one had a conclusive explanation. Theories in profusion have been offered by observers. Some are plausible; others would be considered "far out." None have really sufficed.

One theory is that a localized severe weather aberration knocked the planes into the Atlantic. Counterarguments say violent weather wasn't stressed by Flight 19 and the search plane; there was an absence of wreckage, oil slicks, and bodies; and the aircraft could have followed the sun to find land.

A magnetic phenomenon might have rendered their compasses useless, causing them to fly in widening circles at sea until they ran out of fuel and went into the ocean. Magnetic aberration was considered a possibility to be investigated. Counterarguments are that many aircraft have crossed and recrossed the region without reporting compass trouble; and again, the absence of flotsam, etc.

A wholesale midair collision could have involved five or six planes. This seems extremely unlikely, and there certainly would have been some debris.

A localized sea phenomenon is possible, although undeterminable, in light of the flight leader's quoted report that "Even the sea doesn't look right." Waterspouts were considered but ruled out, since they would be most apt to affect low-flying aircraft, and it wasn't likely that they would erase six planes without leaving clues. Besides, they could have been seen at a distance and avoided.

Capture by UFOs sounds far out, but when you come right down to it we really do not know if it's impossible. Even so, capture of six aircraft seems like a big project.

Capture by as yet unknown people residing beneath the sea and coming to the surface in their vehicles deserves the same comment here as did the previous point.

In short, the disappearance of Flight 19 and the Mariner search plane is still an unsolved mystery.

Other Aircraft Disappearances

Here are but some of many, each with no clues as to the causes.

In 1947 a U.S. flying superfortress winged into nothingness about a hundred miles from Bermuda.

Now classic is the case of the *Star Tiger*, a four-engine Tudor IV passenger aircraft operated by British-South American Airways, which disappeared with twenty-three passengers and a crew of six some 400 miles northeast of Bermuda on January 29, 1948. Weather was good, and her pilot had reported excellent performance.

Three days after Christmas, 1948, a chartered DC-3 became airborne at San Juan, Puerto Rico, and headed for Miami with thirty-two passengers. Flying conditions couldn't have been better. Approaching Florida's coast, the airliner's captain radioed the Miami Airport control tower that they were only fifty miles out, everything was fine, and they would be landing soon. That was the last ever heard from or about the DC-3. It literally vanished into nothingness within sight of Miami.

January 17th, 1949, BOAC's *Star Ariel*, a sister aircraft of the *Star Tiger*, commanded by a veteran transoceanic skipper, winged into limbo during a four-hour trip from Bermuda to Kingston, Jamaica. Weather and sea conditions were good, as reported by the pilot in a final message which positioned the planes approximately 180 miles south of Bermuda. The toll: twenty passengers and crew.

Less than a month separated the two foregoing disappearances. A large-scale sea-air search by the U.S. Navy, U.S. Coast Guard, British planes and vessels, and volunteer merchant ships failed to uncover a trace of the *Star Ariel*.

Among the many aircraft disappearances now on the Bermuda Triangle's lengthening list are these: (*1*) A British York, with a crew of six and thirty-three passengers, February 2, 1953. There was a terse distress call, but it furnished no clues. (2) A U.S. Navy patrol bomber, November 9, 1956. No radio messages. (3) In April of 1962 a twin-engine private plane disappeared on a flight to Nassau, Bahamas. (4) A U.S. Air Force KB-50J refueling tanker plane, en route from Langley Air Force Base, Virginia, to the Azores with nine men. The aircraft's commander reported their position when about 240 miles east of the Virginia coast. After that, silence. A five-day search by vessels and from aloft turned up nothing. Weather was believed to have been good in the area where the KB-50J last reported.

(5) A U.S. Navy "Connie"—a Super-Constellation—disappeared in October of 1954 without so much as a peep from its two radio transmitters. Sharing the aircraft's fate were forty-two people, some of them families of Navy men. There was a familiar sequel: An extensive air-sea search found not a single trace. (6) From Fort Lauderdale to Freeport, Grand Bahama Island, is only a hop-skip-and-jump by air, even for a private plane. Yet a Cessna 180 and two passengers went *poof!* en route. No "Mayday!" call; no wreckage anywhere along the pilot's announced flight path. Date of that "black magic" was June 1, 1973.

(7) Then there are cases that start out as mysteries, but are solved—or at least satisfactorily explained.

August 28, 1963, brought a doubleheader. Two U.S. Air Force KC-135A Strato-tankers, their crews totaling eleven men, took off from Homestead Air Force Base in Florida on a routine mission out over the Atlantic: their assignment, to rendezvous with and refuel two Air Force long-range bombers. Nicknamed "flying gas stations," the KC-135s were designed for in-flight refueling of other military aircraft. They were a kind of military version of the Boeing 707 airliner, each thrust by four powerful jet engines. They could cruise at 500-600 m.p.h. and had a range of about 4,500 miles.

A routine radio communiqué from the mission at noon placed the two tankers 900 miles northeast of Miami, 300 miles southwest of Bermuda. Their ETA back in Homestead AFB was 2:00 P.M. Nothing further was heard from the KC-135s, and when they had not returned by 3:00 P.M., a massive surface and air search was launched. Two strange aspects of the Strato-tankers' disappearance were the good flying weather in their area and the abrupt cessation of radio transmissions.

Since weather was good and the aircraft could not have run out of fuel, the possibility of a midair collision immediately suggested itself. This was given support after a couple of days by the sighting of a large oil slick by one of the search planes and by discovery of a large quantity of debris by the freighter *Azalea City*, both findings in the general area where the KC-135s last reported their position, roughly 300-350 miles southwest of Bermuda. In the flotsam retrieved by the freighter were inflatable life rafts and jackets of the type carried by the missing planes and, most significantly, a flight helmet carrying the stenciled name "Gardner." One of the KC-135's men was Captain Gerald Gardner of Lincoln, Nebraska.

That seemed to be that. Then a new note of mystery injected itself with the sighting from aloft of a second concentration of flotsam 160 miles from the first. If the Strato-tankers had collided and exploded in midair, how come the two patches of flotsam were so far apart? If the second concentration was from one of the KC-135s, it indicated that each had run into its own brand of trouble and had gone into the Atlantic separately. The odds against this happening to two such sophisticated aircraft, in favorable flying weather and without either issuing a distress call, are so astronomically high as to practically rule it out entirely. And the clincher came when a Coast Guard vessel found that the second patch of flotsam consisted of driftwood, seaweed, and a few miscellaneous junk items—nothing from either of the KC-135s.

The mysterious angles cleared up, solution of the case of the multimillion-dollar Strato-tankers appeared clear and crisp: Collision and explosion in midair. There's one question that never will be answered, however: How did it happen?

Now, Other Views

In researching the Bermuda Triangle I was interested in what the U.S. Coast Guard thinks of this seemingly malevolent region and the numerous calamities attributed to it. Among the Coast Guard's major functions is the protection of life and property at sea, which naturally involves search and rescue operations. These functions have brought the Coast Guard into extremely intimate association with the Bermuda Triangle region for a long time. I figured that the USCG's opinion should be valued highly, so I contacted national headquarters in Washington and asked for it. As usual, in numerous contacts I've had with the Coast Guard as a writer in the past, the reply was prompt and detailed. For it I want to thank Commander J. C. Goldthorpe, Acting Chief, Public Affairs Division, as well as Ensign Kathy Kiely for her assistance.

Cmdr. Goldthorpe sent along the Coast Guard's consideration of the region. In his accompanying letter he prefaced it with a significant statement: "The scarcity of Coast Guard . . . research concerning the Bermuda Triangle reflects this area's *lack of abnormal prominence* in Coast Guard operations" (italics mine).

And here is the USCG's opinion, verbatim:

The "Bermuda Triangle" or "Devil's Triangle" is a *mythical* geographic area located off the southeastern coast of the United States. It is noted for an apparent high incidence of unexplained losses of ships, small boats and aircraft.

The Coast Guard does not recognize the existence of the so-called "Bermuda Triangle" as a geographic area of specific hazard to ships or planes. There has been nothing discovered, in review of many aircraft and vessel losses in the so-called "triangle" area over the years, that would indicate that the casualties were the result of anything other than physical causes. No extraordinary factors have ever been identified.

The very location of the so-called "triangle" is probably the greatest single factor responsible for the relatively high losses in the western Atlantic over the years. Although the area is usually described as a triangle extending from Miami to the island of Puerto Rico, along the sweep of the Bahama Banks, thence to Bermuda and back to Miami, the actual area of higher-than-normal search and rescue activity is probably better described as a semicircle extending from the area previously described to Cape Sable, Nova Scotia, then coastal to Miami.

Within the area described above converge the ocean currents known as the Antilles Current, the Florida Current, the North Equatorial Current (all three combine to make up the Gulf Stream) and the Labrador Current. The first three great rivers of water within the Atlantic are considerably warmer than the surrounding water, while the latter (*Labrador Current*) is much colder. The interface of these thermal energy-transfer conduits within the continental land and air mass sometimes creates micro-weather systems of small size, having great energy and violence which can and do inflict serious damage to vessels or aircraft caught up in them. The microweather factor is probably the principal cause of vessel and aircraft losses within the area.

Other factors, of course, are involved. These include simple incompetence on the part of the boat operator or aircraft pilot, mechanical failures which result in sudden catastrophic accidents, collision and the running-down of small craft by larger merchant vessels during darkness or in heavy weather.

Perhaps the most controversial and least documented of all these additional factors is that of piracy or hijacking.

The area encompassed within the semicircle is heavily traveled by surface vessels and aircraft, as it includes within it highly attractive recreational areas, busy commercial trade routes, and rich fishery grounds. These factors alone greatly increase the possibility of increased accidents, simply by reasons of a greater-than-normal number of vessels and aircraft in transit within the area. This, of course, amplifies the probability that one or a combination of the other factors previously identified may come into play in an incident situation.

Contemporary accounts of "The Bermuda Triangle Mystery" make frequent references to magnetic differences in the region which cause radical compass reaction and other physical effects. The Coast Guard has been unable to document a single case of such nature. Nor has a review of the records of the other armed forces, which frequent the air and waters of the world, revealed an account of such a phenomenon.

The Coast Guard's opinion concludes with a select bibliography citing the following: "The Great Bermuda Triangle Rip-Off" by Philip J. Klass, *Pastimes* magazine (Eastern Airlines), July, 1975; *The Bermuda Triangle Mystery—Solved* by Lawrence David Kusche; "Neutercaines" by Frank G. McGuire in *Oceans* magazine, January, 1975; and "Lost Patrol" by Michael McDonnell, *Naval Aviation News* for June, 1973.

"What about scientific opinions?" you ask.

Well, several acknowledged experts in the fields of oceanography, meteorology, and what we might call marine physics have expressed views on the region now popularly known as the Bermuda Triangle. Among those who have been quoted is my friend Dr. F. G. Walton Smith, internationally esteemed oceanographer and leading authority on the Gulf Stream (which flows through the Triangle). Since the opinions of this particular group of scientists agree on major points, they can be presented here as a consensus.

In essence this consensus is in accord with the U.S. Coast Guard's opinion, quoted earlier. That is, there's no mystery about the disappearances of vessels and aircraft within the so-called Bermuda Triangle. Rather, they're caused by recognizable factors —those mentioned by the Coast Guard, operating singly or in combination. Scientific opinion accents adverse weather and sea conditions, which include violent turbulence for planes, known to occur in the region, often within a relatively small area and without

251

appreciable warning. Also recorded for the region are more widespread storms severe enough to place larger vessels and experienced crews in jeopardy. Conceivably higher-altitude weather conditions pose problems for large aircraft. On record—and not only for the Bermuda Triangle—is turbulence severe enough to toss around even the biggest jets.

Like the Coast Guard when stressing the weather factor, scientists point out that insufficiently experienced small-craft skippers—perhaps some amateur aircraft pilots too—cannot or do not interpret and *heed* weather warning signs, and also are unable to cope with adverse conditions when they do occur. It should be added that many pleasure boaters *everywhere* create potentially dangerous situations for themselves by infractions of laws and common sense and by failing to recognize their own and their boats' limitations. Further, there are all sorts of other complications, such as mechanical difficulties and errors in judgment. When these factors are coupled with the increasing numbers of pleasure boats and private aircraft, the lengthening list of casualties within the Bermuda Triangle loses much of its mystery.

And there are logical explanations for absence of debris and bodies, as well as for a lack of distress calls—at least in some instances.

The broad sweep of the Gulf Stream and other currents can scatter wreckage, telltale debris, bodies and survivors, dispersing the evidence in the Atlantic's vastness. So can the conditions that caused the calamities in the first place. And it's a gruesome fact that sharks and other fishes can eliminate bodies and survivors.

Lack of distress calls can be explained by the proven possibility that even vessels of appreciable size can sink so fast that there is no time for a radio message. On record are cases of ships which sank within minutes after being struck by unusually fierce storms or sudden, freakish waves of great height or winds of incredible velocity. Supposing, just for the sake of argument, a radio operator happened to be away from his shack at the critical moment. Or maybe he was thrown against a bulkhead and knocked cold.

Scientists are realists. Although they keep an open mind, they must go with established facts. That is why those whose opinions I've read cannot yet hold with theories of mysterious forces—gigantic vortexes or whirlpools sucking vessels down to their doom, powerful magnetic fields causing navigation instruments to behave erratically and possibly even stall aircraft engines, and so on—at work in the Bermuda Triangle.

Nor are the Coast Guard and scientists ready to buy—publicly, anyway—the theory that UFOs are involved. Scientists are among the first to admit that nothing is impossible; but, even if they were to subscribe personally to the existence of UFOs, they probably would tell you that there has been nothing yet to link such vehicles with incidents in the Bermuda Triangle. The same goes for the theory of possible sea-to-air UFOs. Sightings of UFOs are reported at sea, but this question is asked, why aren't there disappearances of automobiles and other land vehicles and their occupants in areas where sightings have been much more numerous?

Disappearances of people from vessels for no apparent reasons seem to be something else again. In cases of one or two individuals, logical explanations include falling overboard in one kind of accident or another. Conceivably, three or four individuals could be swept away by heavy weather or might even be knocked overboard by a collision at night or in fog. Or some emergency could prompt hasty abandonment in panic. When entire crews are involved, however, satisfactory explanations often become very elusive. Supernatural occurrences and UFOs are convenient answers—and may be correct, for all we know; but at the present such "solutions" are like replying to a question with another query for which there is no answer. Even if we presuppose the existence of vehicles from outer space or under the sea, we have nothing more substantial— yet—than conjectures with which to link them with the disappearances. For the time being, at least, several of these Bermuda Triangle occurrences *are* mysteries.

There You Have It

Two different schools of investigative thought are camped alongside each other on the shores of the Bermuda Triangle. One views the region's goings-on as baffling mysteries that defy solution by existing knowledge. In the other camp are logical proposals by the U.S. Coast Guard and scientists. Facing both camps is the fact that millions of passenger-miles are logged annually by vessels and aircraft without incident.

Which faction to join?

What can I tell you? I'm a visitor here myself. Personally, I can see things going for both schools of thought. If one is favored over another, I guess it boils down to whether an observer is inclined to be an imaginative romanticist or a practical realist.

I'm trying to be both by keeping an open mind.

31

More Mysteries in the Making?

Chances are, you won't see a publication titled *Notice to Mariners* unless you're a professional seafarer or, maybe, a pleasure boat skipper of long standing. It's a technical publication prepared specifically to assist United States vessels in navigation all over the world, with an accent on safety and coverage encompassing every body of salt water you could name—plus many you couldn't. The Defense Mapping Hydrographic Center in Washington, D.C. publishes *Notice to Mariners*, and it's compiled jointly with the National Ocean Survey and U.S. Coast Guard.

The publication carries a wide assortment of items. There are advisory bulletins apprising users of such important details as wrecks and derelicts menacing shipping, changes in buoyage, dangerous reefs discovered and treacherous shoals that have built up, alterations in harbor approaches, discontinuation of lights and other aids to navigation, and so on, even earthquake activity that might threaten shipping.

Also in *Notice to Mariners* are intriguing little items that can leave a reader wondering if perhaps there are mysteries of the sea—or at least potential enigmas—that are not publicized beyond this government publication. Invariably the bulletins are all the more haunting because of their terseness. What follow are samples, reproduced as they appeared in various editions.* Those items with an asterisk after their location may involve the Bermuda Triangle. All are offered here to stimulate your imagination . . . and, I hope, make you want to read more about the sea.

WEST INDIES, VIRGIN ISLANDS*
Venezuelan M/V ANGELA possibly lost and disabled with sick woman, vicinity of Virgin Islands or Anegada Passage. Vessel is gray and black with white 15 by 4 foot flag, length unknown.

---

*It's omitted here to avoid monotonous repetition, but in each case when a vessel is overdue or missing, or it is thought there may be survivors of a calamity at sea, shipping is asked to keep an eye open and notify the Coast Guard of findings whenever possible.

Original report received by CB radio Channel 4 before communications lost.

EASTERN PACIFIC
22-foot sailboat abandoned and adrift in 28-26 N., 128-00 W. Vessel has blue hull, white waterline, white cockpit, broken black mast, numbers 35690F on bow.

WESTERN ATLANTIC, CARIBBEAN SEA*
160-foot M/V TRADE WIND/HO9172 is unreported on voyage from Miami to Margarita I., Venezuela. Vessel has gray hull, white superstructure aft, 10 cars on deck.

ALASKA, ATTU ISLAND
Due to recent earthquake in area of Attu Island, it has been reported that the bottom of Massacre Bay has risen approx. 4 to 7 feet. Mariners are urged to use caution in area.

WEST INDIES*
41-foot white and blue sailing vessel NAN-PRI overdue on voyage from Key West, Fla., to Tortola, British Virgin Islands.

BAHAMA ISLANDS*
The 60-foot houseboat GLORY MINE, previously reported sunk, has been reported awash in 21-47N., 74-15 W.

NORTHWEST PACIFIC
508-foot Liberian vessel GERANIUM/5MYZ reported missing. Last reported position, 50.6 N., 178.5 E., en route to Osaka, Japan. Vessel is bulk carrier with logs on deck.

WINDWARD ISLANDS, CARIBBEAN*
CESSNA 172 aircraft, green and white, 3 persons on board, reported overdue on flight from St. Croix to St. Lucia via Martinique. Plane possibly ditched in water.

Here's one that poses interesting possibilities:
NORTH PACIFIC
Derelict 4-ton raft, constructed of 32 steel drums, 2 masts, schooner rigged, 42 feet long, 6 feet wide, reported adrift in 29-37 N., 171-44 W.

About the above we might ask, is somebody trying to imitate Thor Heyerdahl and the *Kon-Tiki*?

PORTUGAL
    200,000-ton vessel suffered considerable damage striking un-
    identified object in 37-55 N., 09-21 W.

That's the way the *Notice to Mariners* read, "200,000-ton."
With that weight she sounds like a supertanker. What makes the
item especially intriguing is that the collision occurred in deep water
on open sea. What kind of object would cause considerable damage
to a ship that size? No mention is made of an errant mine from
World War II.

PHILIPPINE SEA
    Timber carrier KIHO MARU sunk in 19-53 N., 125-42 E. 20
    persons still missing. Timbers adrift in 20-13 N., 125-18 E.

CARIBBEAN SEA*
    Underwater volcanic eruption reported in 12-18 N., 61-30.5 W.

WEST INDIES*
    43-foot Columbia sloop JOHN GALT with 4 or 5 persons on
    board, unreported on trip from San Juan to Fort Lauderdale, Fla.
    Vessel has white flush deck, white hull with blue rim, teak trim,
    sail No. 43, gray zodiac on davits, call sign WY8976.

EAST AFRICA
    Merchant ship reported receiving small arms fire from uniden-
    tified single-engine plane close offshore near border of Mozam-
    bique and Republic of South Africa.

Our sport fishing boat enjoyed much nicer hospitality when I
looked for marlin in the Mozambique Current in that area a while
back.

HAWAIIAN ISLANDS AREA
    63-foot white hulk, white with orange superstructure, adrift in
    23-20 N., 160-15 W.

A doubleheader in the Bermuda Triangle region?
NORTHWESTERN ATLANTIC
    41-foot catamaran SEA HORSE IV with 3 persons on board
    reported overdue from Annapolis, Md., to Marsh Harbor,
    Bahamas. Vessel has white hull and decks, red bottom, letter A
    on mainsail and red/yellow/blue spinnaker.
    40-foot catamaran previously reported capsized and adrift off
    coast of New Jersey and Delaware is believed to be in vicinity of

37 deg. N., 64 deg. W. If sighted, vessels are requested to make all efforts to identify by obtaining name from bow or transom. A yellow or orange 4-man life raft with 3 persons on board is believed adrift from the catamaran.

On February 18, 1974 in Harrisburg, Pennsylvania, Thomas L. Gatch of Virginia began his attempt to be the first person to cross the Atlantic Ocean in a balloon. His equipment included a special, pressurized gondola of impact-resistant fiberglass with built-in flotation to keep it upright and afloat if the balloon went down. He estimated his crossing time at five days and expected to land somewhere in southern France or northern Spain. An airliner spotted the balloon when about 1,300 miles due east of Miami. Two days later a Spanish freighter reported sighting it, farther along. At that time, however, it appeared that the balloon was in the grip of air currents that were shoving it much farther south than its one-man crew had anticipated. Gatch didn't make it. His balloon disappeared, and on February 23, it was calculated that he had gone into the water in an area some 800 miles south-southwest of the Azores.

The March 30, 1974 *Notice to Mariners* ran the following item to request shipping to keep watching:

NORTH ATLANTIC

Missing balloonist Thomas L. Gatch was flying an eight balloon cluster, black in color, with white gondola, and a red banner below that. He was . . . last sighted 21 February. A vessel reported on 9 March: Sighting object, unconfirmed, resembling gondola on surface 04-14 N., 12-15 W.

On March 6 the U.S. Department of Defense called off an air-sea search that had combed nearly 225,000 square miles of Atlantic. It failed to find any trace of the balloon or its gondola. Some observers have since wondered, was the intrepid Thomas L. Gatch a Bermuda Triangle casualty or mystery?

EASTERN PACIFIC

S/V VALHALLA/WJZ3751, 45 feet, natural wood hull, white cabin, white sails, with 3 persons on board, en route Hawaii to San Francisco, California, reported out of fuel and water. Last position, 27-00 N., 127-00 W. Attempting to make Mazatlan, Mexico. Radio frequencies 2182/2638 kHz.

WESTERN CARIBBEAN*

Motor vessel BERNICE M. overdue Puerto Barrios, Guatemala, to Santo Domingo, Dominican Republic. BERNICE M. is 225-foot cargo vessel with gray hull and white cabin aft.

FRENCH ANTILLES, GUADELOUPE (Caribbean)

Because of threats of eruption of the volcano La Sourfriere, navigation is prohibited in the area [outlined]. Navigation is considered dangerous within 5 nautical miles of the shores of the south part of Basse-Terre from Anse a la Barque to Capesterre.

STRAITS OF FLORIDA*

56-foot landing craft, all gray, 3 persons on board, reported overdue Fort Lauderdale to Freeport, Grand Bahama.

CARIBBEAN SEA

54-foot dismasted S/V FEISTY reported adrift in approx. 12-39 N., 72-06 W.

And in the same general expanse:

An unlighted derelict cabin cruiser sighted adrift in 13-45 N., 77-40 W.

EASTERN PACIFIC

S/V DRUM overdue on voyage from Hawaii to Port Angeles, Washington. Vessel is 25-foot trimaran, white hull, gold trim, gold cabins fore and aft, white deck, 2 persons on board.

Of such stuff—transoceanic crossings in small sailboats—are tragedies, not mysteries made.

Another brave—or foolhardy—soul:

NORTH ATLANTIC

The single-handed trimaran THREE CHEERS has been reported overdue on voyage from England to Rhode Island. Trimaran has large number 99 painted on hull and deck.

NORTH CAROLINA TO VIRGIN ISLANDS*

30-foot S/V ARK, double ended sloop rig, white hull and cabin, name on sail, 4 persons on board, unreported on voyage Morehead City, NC, to St. Thomas, VI.

One month—it happened to be August—was busy in California. A single issue of *Notice to Mariners* listed no less than five boats overdue on relatively short trips: The 30-foot cruiser *Aqua-Gem* with four persons aboard; a 39-foot fishing vessel *Island Clipper* (presumably a public sport fishing boat), with ten to eighteen on board; a 25-foot cabin cruiser, five aboard; a 15-foot outboard, with five persons, in the ocean; and a 40-foot catamaran *Aloha* (a prophetic name, maybe), with at least six people aboard.

JAPAN, OGASAWARA GUNTO, NISHINO SHIMA

New islet observed in 27-14.6 N., 140-52.9 E., named Nisino Sima Sinto, has length about 750 meters east to west, width about 400 meters north to south, and height about 40 meters above surface. Volcanic eruption is continuously going on.

WESTERN ATLANTIC

41-foot sloop SPRAY, white hull, broken mast, abandoned in 33-01 N., 70-25 W.

VIRGINIA, OFFSHORE

Small aircraft crashed and sank in approximate position 36-52 N., 75-34 W.

HAWAIIAN ISLANDS

S/V GITANA DELMAR, 42-foot trimaran, plywood and fiberglass construction, white hull, blut bottom, sand colored decks, one person on board, reported sinking in 21-15 N., 163-00 W. Vessell en route Kwajalein Atoll to Honolulu.

A double, or a triple, for the Bermuda Triangle?

NORTH ATLANTIC

The 60-foot yacht SHAMROCK, white hull with blue bottom, reported adrift in 35-30 N., 67-45 W.

The 41-foot ketch ANGELICA, white and green, and the 35-foot trimaran MERIDIAN reported overdue between Bermuda and eastern U.S.

PERSIAN GULF AREA

Ships transiting the Persian Gulf area, particularly in the Strait of Hormuz, should be alert to unusual or abnormal activities that could lead to a hijacking or other hostile actions.

CARIBBEAN SEA

38-foot Salvage vessel NUMA with 2 persons on board, unreported from Grand Cayman Island since 1 July [item was dated 13 July]. Vessel has gray steel hull, white cabin forward, air compressor and diving platform aft.

From the description of the missing vessel it sounds like a treasure-hunting expedition.

EASTERN PACIFIC

32-foot vessel VILIA reported overdue [nearly two weeks late at time of bulletin] on voyage from Balboa, Canal Zone, to San

Diego. Vessel has white hull, blue stripe, teak decks, blue cabin top. Call sign, WYZ5507. Two persons aboard or in inflatable rubber raft or 7-foot fiberglass dinghy.

Finally, a few items with spicy possibilities:

GULF OF MEXICO

On 5 August [1973] Panamanian vessel PERSEUS and Mexican vessel PUEBLA collided 90 miles off Yucatan, resulting in the loss of 390 steel drums of potassium cyanide and sodium cyanide, which are extremely poisonous.

Each drum was two and one-half feet high and fourteen to sixteen inches in diameter and contained 100 pounds of cyanide (which is the chemical used in gas chambers for executions). Total amount of the deadly poisons lost came to 39,000 pounds, or nearly twenty tons.

EASTERN ATLANTIC

Cylindrical object, 3 feet diameter, sighted in 34-22 N., 20-19 W. Believed to be floating mine. [Contact with it can confirm.]

SAN PEDRO CHANNEL (California)

Torpedo lost in vicinity 33-32 N., 118-10 W. since 30 April (*then about two weeks*). Torpedo is 104 inches long, 12 1/2 inches diameter, painted black nose, orange/white/gold midsection, red propeller.

There have been many, many items concerning overdue, unreported, derelict, possibly missing vessels in *Notice to Mariners*. It's not to be inferred, however, that all these bulletins represent mysteries in the manner of the *President, Atalanta, Mary Celeste, Cyclops, James B. Chester*, and other enigmas detailed elsewhere in this book. Many of the vessels variously listed as overdue or unreported eventually turn up in some port or are sighted and assisted by merchant ships or the U.S. Coast Guard or U.S. Navy. Or, if they do not show or are not found, the reasons can be determined. And there are probably ordinary explanations for numerous abandoned vessels found derelict, such as breaking loose from moorings and drifting out to sea.

Nevertheless, among the bulletins also exists a potential for occurrences which may never be explained satisfactorily. There is graphic proof not only in the Bermuda Triangle, but also on waters all over the world.

Consider this eye-riveting excerpt from an article by F. W.

Fricker, National Information Services, Defense Mapping Hydrographic Center:

"It has been reliably reported that during the ten-year period 1962-1972, 71 vessels with 1,034 men, women and children aboard sailed out of ports all over the world and were never seen again."

32

There Will Be More Stories

We've come to the end of this voyage. But for as long as man and the sea continue their association—a relationship that will endure while there is human life on this globe—they will together write an endless procession of absorbing stories.

Selected Bibliography

These are fine books, highly recommended for further reading. Dates shown are the latest known to the author, but in some instances there may have been later printings.

A Pictorial History of Sea Monsters and Other Dangerous Marine Life, James B. Sweeney; Crown Publishers, Inc., New York, 1972.

Amelia Earhart Lives, Joe Klaas; McGraw-Hill Book Co., New York, 1970.

Astounding Tales of the Sea, Edward Rowe Snow; Dodd, Mead & Co., New York, 1965.

Captain Joshua Slocum, Victor Slocum; Sheridan House, New York, 1950.

Halliburton, the Magnificent Myth, Jonathan Root; Coward-McCann, Inc., New York, 1965.

Invisible Horizons, Vincent Gaddis; Chilton Book Co., Radnor, Pa., 1965.

Invisible Residents, Ivan T. Sanderson; The World Publishing Co., New York, Cleveland, 1971.

Life magazine (*Joyita* story), December 12, 1955.

Limbo of the Lost—Today (Bermuda Triangle), John Wallace Spencer; Phillips Publishing Co., Westfield, Mass., 1975; and paperback edition, Bantam Books, Inc., New York, 1975.

Men and Ships Around Cape Horn, 1616-1939, Jean Randier (translated from the French by M. W. B. Sanderson); David McKay Co., Inc., New York, 1969.

Mysteries of the Sea, Robert de la Croix; The John Day Co., 1970.

Notice to Mariners, a technical periodical published by Defense Mapping Agency and prepared jointly with the National Ocean Survey and U.S. Coast Guard, Washington, D.C.

Perils of the Port of New York, Jeanette Edwards Rattray; Dodd, Mead & Co., New York, 1973.

Phantoms of the Sea, Raymond Lamont Brown; Tapplinger Publishing Co., New York, 1973.

Richard Halliburton (a book based on his letters to his parents and notes about his life); The Bobbs-Merrill Co., Indianapolis and New York, 1940.

Sea Fights and Shipwrecks, Hanson W. Baldwin; Hanover House, Garden City, N.Y., 1955.

Ship Ashore!, Jeanette Edwards Rattray; Coward-McCann, Inc., New York, 1955.

Sinkings, Salvages and Shipwrecks, Robert F. Burgess; American Heritage Press, New York, 1970.

Some Ship Disasters and Their Causes, K. C. Barnaby; American edition by A. S. Barnes & Co., Cranbury, N.J., 1970.

Supernatural Mysteries and Other Tales, Edward Rowe Snow; Dodd, Mead & Co., New York, 1974.

The Bermuda Triangle, Charles Berlitz; Doubleday & Co., Garden City, N.Y., 1974.

The Bermuda Triangle Mystery—Solved, Lawrence David Kusche; paperback edition by Warner Books, Inc., New York, in arrangement with Harper & Row, 1975.

The Devil's Triangle and its sequel *The Devil's Triangle 2* (both about the Bermuda Triangle), Richard Winer, paperback editions by Bantam Books, New York, 1975.

The Fury of the Seas, Edward Rowe Snow; Dodd, Mead & Co., New York, 1964.

The Great Iron Ship (detailed account of the life of the steamship *Great Eastern*), James Dugan; Harper & Brothers, New York, 1953.

The Morro Castle, Hal Burton; The Viking Press, New York, 1973.

The Ocean River (*the* book about the Gulf Stream), Henry Chapin and F. G. Walton Smith; Charles Scribner's Sons, New York, 1952.

The Search for Amelia Earhart, Fred Goerner; Doubleday & Co., Garden City, N.Y., 1966.

The Seas in Motion, F. G. Walton Smith; Thomas Y. Crowell Co., New York, 1973.

True Tales of Terrible Shipwrecks, Edward Rowe Snow; Dodd, Mead & Co., New York.

Women and Children Last (detailed account of the steamship *Arctic* disaster), Alexander Crosby Brown; G. P. Putnam's Sons, New York, 1961.